T0137346

Springer Theses

Recognizing Outstanding Ph.D. Research

Aims and Scope

The series "Springer Theses" brings together a selection of the very best Ph.D. theses from around the world and across the physical sciences. Nominated and endorsed by two recognized specialists, each published volume has been selected for its scientific excellence and the high impact of its contents for the pertinent field of research. For greater accessibility to non-specialists, the published versions include an extended introduction, as well as a foreword by the student's supervisor explaining the special relevance of the work for the field. As a whole, the series will provide a valuable resource both for newcomers to the research fields described, and for other scientists seeking detailed background information on special questions. Finally, it provides an accredited documentation of the valuable contributions made by today's younger generation of scientists.

Theses are accepted into the series by invited nomination only and must fulfill all of the following criteria

- They must be written in good English.
- The topic should fall within the confines of Chemistry, Physics, Earth Sciences, Engineering and related interdisciplinary fields such as Materials, Nanoscience, Chemical Engineering, Complex Systems and Biophysics.
- The work reported in the thesis must represent a significant scientific advance.
- If the thesis includes previously published material, permission to reproduce this must be gained from the respective copyright holder.
- They must have been examined and passed during the 12 months prior to nomination.
- Each thesis should include a foreword by the supervisor outlining the significance of its content.
- The theses should have a clearly defined structure including an introduction accessible to scientists not expert in that particular field.

More information about this series at http://www.springer.com/series/8790

Zubair Iftikhar

Charge Quantization and Kondo Quantum Criticality in Few-Channel Mesoscopic Circuits

Doctoral Thesis accepted by
the University of Paris-Sud, Orsay, France

Author
Dr. Zubair Iftikhar
University of Paris-Sud
Orsay, France

Supervisor
Dr. Frédéric Pierre
University of Paris-Sud
Orsay, France

ISSN 2190-5053 ISSN 2190-5061 (electronic)
Springer Theses
ISBN 978-3-030-06899-8 ISBN 978-3-319-94685-6 (eBook)
https://doi.org/10.1007/978-3-319-94685-6

Printed on acid-free paper

This Springer imprint is published by the registered company Springer International Publishing AG part of Springer Nature
The registered company address is: Gewerbestrasse 11, 6330 Cham, Switzerland

Supervisor's Foreword

The Ph.D. thesis of Zubair Iftikhar investigates experimentally several facets of the quantum electronic correlations induced by the Coulomb interaction in nanoscale circuits. The two major fundamental problems addressed are the quantized character (the discreteness) of charge in circuits and the intriguing states of matter that can arise in strongly correlated and quantum critical systems. The thesis constitutes a remarkable breakthrough on these long-standing questions, which was made possible by a novel kind of highly tuneable hybrid metal–semiconductor device cooled to record low temperatures.

The first part of the thesis deals with the control of charge quantization in circuits by quantum fluctuations. Whereas the charge in small, weakly connected conductors is quantized in units of the elementary electron charge e, quantum fluctuations progressively reduce charge discreteness as the connection strength is increased. This work demonstrates some of the most fundamental theoretical predictions that had so far eluded experimental confirmations, including the absence of charge quantization in the presence of a ballistic quantum channel, as well as the expected scaling with the distance to the ballistic critical point. The achieved understanding and full on-demand control of charge quantization may lead to a broad range of applications. It has direct consequences regarding devices based on single-charge manipulations (a field of study called 'single electronics'), and the central role of charge quantization in the different quantum laws of electricity with coherent conductors signals direct quantum engineering implications for future nanoelectronics.

The second part and third part of the thesis address the fundamental topic of electron correlations and quantum critical systems, which underpins unconventional behaviors of immense potential such as heavy fermion materials and high Tc superconductors. This thesis sheds light on some of the exotic physics of quantum phase transitions through the precision experimental implementation of a model that exhibits the richest variety of quantum critical points. Two degenerate charge states of a small metallic island constitute the equivalent of the spin-1/2 magnetic impurities in the Kondo model. This charge pseudospin made it possible to investigate the 'multichannel' generalization of the Kondo model, including in uncharted territory. In particular, a highlight of this thesis is the first observation of a genuinely interacting quantum critical state, with irreducibly strong interactions, associated with a renormalization flow of the conductance toward a universal intermediate value.

Marcoussis, France
March 2017

Dr. Frédéric Pierre

Abstract

This thesis explores several fundamental topics in mesoscopic circuitries that incorporates few electronic conduction channels.

The first experiments address the quantized character (the discreteness) of charge in circuits. We demonstrate the charge quantization criterion, observe the predicted charge quantization scaling, and demonstrate a crossover toward a universal behavior as temperature is increased.

The second set of experiments addresses the unconventional quantum critical physics that arises in the multichannel Kondo model. By implementing a Kondo impurity with a pseudospin of $1/2$ constituted by two degenerate charge states of a circuit, we explore the two- and three-channel Kondo physics. At the symmetric quantum critical point, we observe the predicted universal Kondo fixed points, scaling exponents and validate the full numerical renormalization group scaling curves. Away from the quantum critical point, we explore the crossover from quantum criticality: direct visualization of the development of a quantum phase transition, the parameter space for quantum criticality, and universality and scaling behaviors.

Acknowledgements

I have always been a privileged person. When I was a child, my nursery teacher used to carry me in her arms when I was crying because my mother left me at school. At home, whereas my brothers used to have fixed times to do their homework, I was free to work or not. In 7th grade, I was in the seaside class. In 8th grade, I had a French teacher, Ms. Galin, who was training us for the *agrégation* together with her. At high school, I had a wonderful maths professor, Mr. Squalli, to whom one could even ask 'but why is it forbidden to divide by zero?'. After the high school also, I had the best professors. In master's, I have been able to make an unforgettable internship in Siberia.

I grew up in the best environments, with the best equipment and with the best people. I had the best friends in the world. Let me greet some of them I still meet sometimes, by chronological order: Wilfried C. who used to share his snake in elementary school; Florian B. who was making fun every time; Alexia H. who used to get angry with everyone but me; Laurent R. with whom I had long philosophical arguments in high school; Alexandre H. who I was happy to see on my Ph.D. defense; Ibrahim who I see at the university library (good luck); Jérémy D., Thomas N., Maxime J., Jérôme D., my friends in *classes préparatoires*, thanks for the good time (the oral trainings, the discussions at the cafeteria, etc.); Hélène S., Romain J., Arthur L., Dorian G., my friends at SupOptique; Jérémie D. and all our '*tutorés*' high school students with whom I had beautiful discussions before, during, and after the mentoring sessions; Hamit M., Nicolas M., Charles P., Ana B., Maria O., Anna I. and all my friends in Russia.

And after Russia, I landed in Marcoussis, at LPN (now called C2N). There, I have a lot of people to acknowledge—first of all, my Ph.D. supervisor, Fred. Once I told him I have nothing to do, he then answered 'You should never be bored!'. I have met some desperate Ph.D. students frustrated by a lack of management, and today I understand I had the best supervisor. I have spent three years at a crazy place. Thanks to Sébastien Jezouin the former Ph.D. student, my colleague during my first year, he taught me many things, e.g., to run to catch up the last bus at 20:09 just after launching the night measurement. Thanks to François Parmentier for teaching me how are working the lab and the experimental setup you are a model even in the way you talk. Thanks to Anne Anthore, because of the kindness of your answers, I have never felt ashamed in asking my questions (often stupid, sometimes deep). Thanks to the remaining team, in particular Ulf Gennser, for your kindness and for rereading my thesis written in English, Yong Jin for the discussions at the cafeteria, and Julien Chaste to have always been there to talk about physics or life. I acknowledge the non-permanent members of the group Amina, Khalifa, Hugo, and Debora. Many thanks also to the other students and postdocs at LPN, in particular those who used to take the bus late in the evening and nicknamed the *habbs* by Riadh: Riadh himself, Shayma, Avishek, Hakim, and the others. Special thanks to Avishek for the beautiful moments aboard the Titanic. And for the noontime breathless football matches on Tuesday, many thanks to Stéphane, Kamel, Lorenzo, Rolland, Paul, Dominique, Khalifa, Lyas, Mathias, Ivens, Juan, Vivek, Daniel, Amadeo. Whether it rains, it snows, or the sun shines, whether we were twelve or only eight, it has always been funny games! Thanks to the students of the optics group Mickaël, Thomas, Fayçal, and Benoît for the good ambiance at the second floor of D2 building.

Thanks to Christopher Bäuerle and Julia Meyer, who accepted to report my thesis. Thanks to Takis Kontos and Serge Florens for their questions during the defense. And thanks to Cristián Urbina for being the president of my Jury and also for the beautiful discussion we had after the defense. I acknowledge also Gilles Montambaux, head of the doctoral school, for reassuring me during the thesis writing time.

Finally, I would like to acknowledge my family, in particular my mother, the author of this thesis; my sister for supporting me with her jokes; and my brother Tahir, because he opened the way to higher studies by showing me that I can succeed.

I wondered about many deep questions all along my life and during the three last years in particular. I experienced relatively hard times, as everyone. I have thought, and I believe I have found the way: Life is a walk which only goal is to love it for itself. I have reached this goal, I like my walk. This explains why I feel privileged, why I am privileged. I would like to thank all the people I have met on my way. The one that goes with me, hand in hand, and who changed my walk in a ballad. And finally, my mother, who put me in this adventure twenty-six years ago and to whom this thesis is dedicated.

Contents

Preliminary Remarks

We use the acronyms defined in Table 1 and the symbol '$\triangleq$' for the definition physical or mathematical quantities and functions. The definition of some Greek symbols may change[1]; nevertheless, the right definition to use is generally obvious.

Table 1 Definition of the acronyms

Acronym	Definition
2DEG	Two-dimensional electron gas
CFT	Conformal field theory
DCB	Dynamical Coulomb blockade
FL	Fermi liquid
FQHE	Fractional quantum Hall effect
MBE	Molecular beam epitaxy
N-CK	N-channel Kondo (with $N = 1, 2\ldots$)
NFL	Non-fermi liquid
NRG	Numerical renormalization group
QCP	Quantum critical point
QHE	Quantum Hall effect
QPC	Quantum point contact
QPT	Quantum phase transition
SET	Single electron transistor
STM	Scanning tunneling microscope

[1]For example, γ is generally $\gamma \triangleq expC$, with $C \approx 0.5772$ the Euler's constant. But sometimes γ refers to a generic critical exponent.

Chapter 1
Introduction

In the late eighties, with the progress in micro- and nano-fabrication, physicists have been able to make devices where they were able to add/remove a single electron to/from small conductors relatively isolated from the surrounding circuit. This possibility has opened a new field of research called *single-electronics* with promising applications such as sensitive thermometry, electrometry, metrology or refrigeration [1].

In some specific limits, this kind of device, as the one studied in this thesis, can be described relatively easily using a perturbation theory. But in general, the quantum transport of electricity reveals complex and intriguing phenomena where many-body effects play a crucial role. Despite the apparent complexity of the problem, it appears that the solution is sometimes very simple. For instance, the conductance of a 2D electronic system measured at low temperature and under a strong magnetic field is $\nu e^2/h$ where ν is an integer called filling factor, e is the elementary charge and h is the Planck's constant. This phenomenon called 'integer quantum Hall effect' was discovered in 1980 [2]. The fact that this conductance only depends on an integer ν and some fundamental constants is absolutely remarkable! This is the kind of experiments discussed in this thesis: the experimental observation of *simple*[1] results emerging from a problem which is a priori highly non-trivial.

This chapter is organized in three parts. The two first sections would provide a general knowledge to understand this thesis whereas the last one is a summary of this book.

1.1 Quantum Conductors

Quantum mechanics gives to the electrons a wave-particle duality. In clean samples and at low temperatures, inelastic scattering with impurities and vibrational phonons are rare, the electron phase is thus conserved over long distances. Transport prop-

[1] A more physical word would be 'universal', meaning independent of the microscopic details of the sample one is considering.

Z. Iftikhar, *Charge Quantization and Kondo Quantum Criticality in Few-Channel Mesoscopic Circuits*, Springer Theses,
https://doi.org/10.1007/978-3-319-94685-6_1

erties will change due to this phase coherence giving rise to observable quantum phenomena. This field of research started in the eighties [3, 4] is named *mesoscopic physics*.

In this section we focus on quantum conductors and the quantum point contact (QPC) in particular. Indeed, this is the basic building block of our quantum circuits. We are able to put such a conductor inside a circuit just as a usual electrical dipole. However, its behavior is quite singular. First, its conductance (the invert of the electrical resistance) shows steps quantized in units of the quantum of conductance $G_K \triangleq e^2/h$, when plotted versus a gate voltage that controls the size of the QPC. Second, its conductance between two steps depends on the other elements of the circuit (e.g. classical resistors or other quantum conductors).

The first subsection introduces the quantum of conductance and the notion of electronic conduction channels. The second one presents our experimental realization of quantum conductors.

1.1.1 Quantum of Conductance

A practical way to describe the transport of electrons in a sample with well localized scatterers has been first proposed by Landauer [5]. It has been refined [6, 7], in particular after experimental observations [3, 8, 9] to eventually give the Landauer-Büttiker formalism.

In the first paragraph, we consider an ideal one-dimensional conductor connecting two electronic reservoirs at thermodynamic equilibrium. In the second paragraph, we consider a more practical situation and discuss the Landauer-Büttiker formalism.

Dimensionality

Let us consider two reservoirs (the left and the right) connected by a pure ideal 1D conductor. The electrical current through the conductor is given by the elementary charge e of an electron times the number of electrons that crosses a section of this conductor per unit of time: $I = e \times \int \frac{\mathrm{d}k}{2\pi} v(k) f(k)$, where k is the (scalar) wave vector associated to an electron, $v(k)$ is the velocity of an electron carrying a momentum k and $f(k)$ counts the number electrons that carry a momentum k [4]. Assuming free propagation, the energy E reduces to the kinetic energy, thus $\partial E/\partial k = \hbar v$; then the velocity simplifies and it comes:

$$I_L = \frac{e}{h} \int \mathrm{d}E f_L(E) \tag{1.1}$$

where I_L is the current injected by the left reservoir. For an electronic reservoir at equilibrium (at a temperature T and a chemical potential μ_L), $f_L(E) = \{\exp((E - \mu_L)/k_B T) + 1\}^{-1}$ is given by the Fermi-Dirac distribution. A net current I will flow when one sets a voltage bias V between the left and the right reservoirs ($eV = \mu_L - \mu_R$). Whatever the temperature, the conductance $G = I/V$ in the limit $V \longrightarrow 0$ takes the value of the quantum of conductance:

$$\boxed{G_K = \frac{e^2}{h}} \tag{1.2}$$

This is the electric conductance of an ideal single-mode conductor. It is limited to the universal value $G_K \approx 1/25.8\,\mathrm{k}\Omega$ by the fermionic nature of the charge carriers. In practice, one may consider non-ideal conductors (embedding scatterers or with a finite width that allow several electronic modes to pass). These situations are discussed in the next paragraph.

Landauer-Büttiker formalism

When a quantum conductor contains defects or irregularities, an incoming electron wavefunction can be either transmitted through the conductor or reflected back to the emitting reservoir. This scattering process is elastic since it is associated to a time-independent potential.

We consider a quantum conductor with a finite width that allows several transverse modes (also called electronic channels). One can build the scattering[2] matrix S that describes the scattering of the incoming waves (from the left and from the right) into the outgoing waves (to the left and to the right reservoirs) [6]. The conductance is then given by the two-terminal Landauer-Büttiker formula at zero temperature:

$$G = \frac{e^2}{h} \sum_n \tau_n \tag{1.3}$$

where the energy dependence of the transmission probabilities τ_n of the electronic channels is assumed to be negligible. Each channel contribute at most to one quantum of conductance e^2/h.

Many channels can be transmitted simultaneously and give rise to a larger conductance. However, as it will be explained later, in our case, the opening of the channels occurs one-by-one (for an illustration, see Fig. 1.3). In the experiments reported in this thesis, we have generally $\tau_1 \leqslant 1$, therefore the $\tau_{n>1} \lll 1$ are negligible.

According to Eq. (1.3), the average current across a quantum conductors is fully characterized by the transmissions τ_n of its electronic channels. Let us see how to build a quantum conductor in practice, and how to control the transmission probabilities τ_n.

[2]In our experiments, we consider a scattering by a smooth potential (a saddle-point constriction).

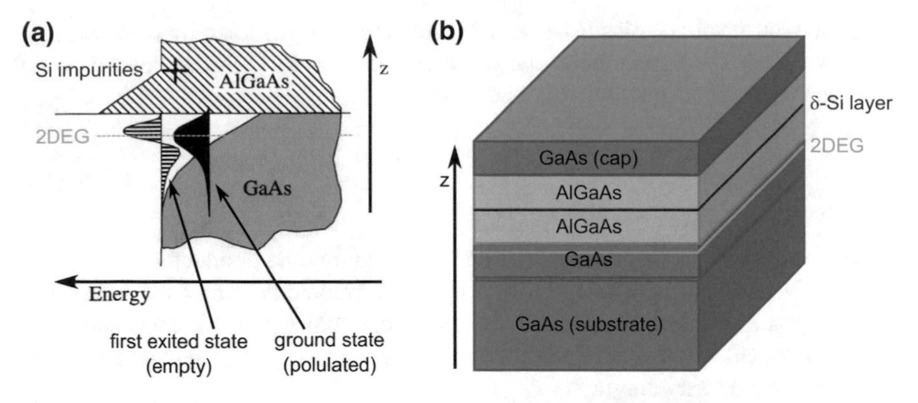

Fig. 1.1 2DEG formed in a GaAs/AlGaAs heterostructure. The z axis shows the direction of the MBE growth. **a** (reproduced from [11]) The electrons given by the silicon impurities (+) fall in the triangular well. The ground and first excited states wavefunctions are represented. At low temperature, only the quantum ground state is populated. The position of these electrons is locked in an (x, y) plane close to the interface GaAs/AlGaAs. **b** Schematic view of our heterostructure. A GaAs cap is used to avoid oxidation. The 2DEG is typically 100 nm below the surface

1.1.2 Quantum Coherent Conductors

Quantum coherent conductors can be fabricated using various methods. Our way is to take a 2D conductor and then to make a constriction of this 2D conductor into a local 1D conductor. That kind of small 1D conductor made in a semiconductor is called a quantum point contact (QPC), we will describe its principle in this subsection.

2D electron gas (2DEG)

The most direct example of 2D electron gas (2DEG) is probably a mono-layer of carbon (graphene). However, the graphene technology is not yet mature to implement QPCs (despite promising solutions [10]).

The 2DEG we have used is made in GaAs using MBE techniques at C2N/LPN by Ulf Gennser, Antonella Cavanna and Abdelkarim Ouerghi. This technique consists in evaporating gallium, arsenic and some amount of aluminum on an initial substrate of GaAs. The crystalline growth is controlled with an accuracy which is better than a single layer deposition. This process is done in an ultra high vacuum chamber to reduce pollution.

One can engineer the conduction band of a stack of GaAs/AlGaAs called heterostructure by modulating the aluminum proportion [11, 12]. The silicon impurities introduced in the AlGaAs region (see Fig. 1.1) ionize. The generated free electrons find the minimum of energy $E(z_{\text{min}})$ and eventually stay there and form the 2DEG.

Since they are trapped in z direction, their motion will be quantized along this axis. When the temperature is lower than the energy spacing between the modes,

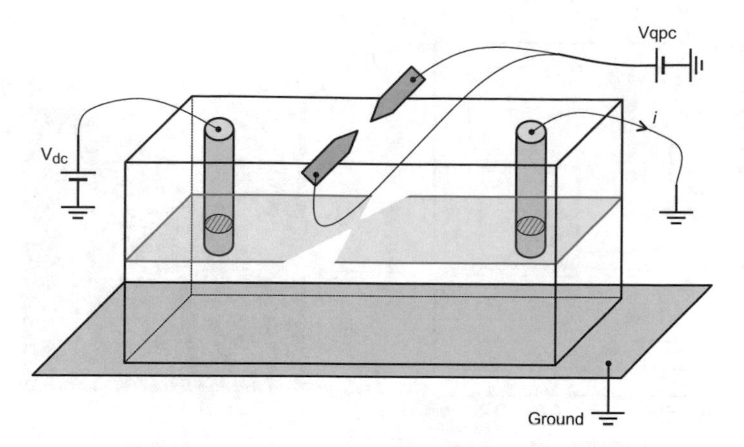

Fig. 1.2 Diagram of a Quantum point contact. The transparent cuboid represents the (Al)GaAs heterostructure. A voltage V_{qpc} applied on the QPC gates generates an electric field between the metallic gates (in grey) and the ground (in gold). This field depletes the 2DEG under the gates. The constriction in the 2DEG is the QPC. The conductance $G = i/V_{\text{dc}}$ of the QPC is measured using electrodes called 'ohmic contacts' (gold cylinders) that connect the 2DEG

only the fundamental level is populated (in black in Fig. 1.1a). They are free to move in the other directions (x, y), but their position in z is locked.[3]

Quantum point contact (QPC)

What we actually need is a 1D conductor. The idea is then to polarize split gates deposited on the top of the (Al)GaAs heterostructure to build a constriction of the 2DEG and eventually get a 1D conductor. This 1D component is called a QPC. The typical gap between the split gates is given by the Fermi wavelength $\lambda_F = \sqrt{2\pi/n} \approx 50\,\text{nm}$, where $n \approx 2.5 \times 10^{11}\text{cm}^{-2}$ is the density of our 2DEG.

The quantum point contact [8, 9] is the basic building block of our mesoscopic circuits. We have three QPCs in our device (see Fig. 1.9). Their working principle is shown in Fig. 1.2. The conductance of a QPC depends on the voltage applied on its two gates. The negatively higher the gate voltage, the narrower the constriction (and thus, the lower the conductance of the QPC). A measurement of the conductance of a QPC with respect to its two gates is shown in Fig. 1.3.

This figure shows pronounced steps at multiples of $G_K \triangleq e^2/h$. At each step, one more electronic channel is transmitted through the QPC. The large plateaus at each step are actually due to the strong perpendicular magnetic field ($B \approx 2.7\,\text{T}$) [13]. In this regime, the channels pass one-by-one. Hence, one can generalize the transmission τ to values larger than a unitary transmission when considering several channels (e.g. $\tau = 3.14$ means that three channels are fully transmitted and a fourth has the intermediate transmission probability of 0.14.)

[3]The ground state wavefunction shown in Fig. 1.1a gives the average position, $\overline{z}$, of the electron in the quantum well. This position can fluctuate, according to Quantum Mechanics, but the motion remains fundamentally two-dimensional, as the average position *cannot* change.

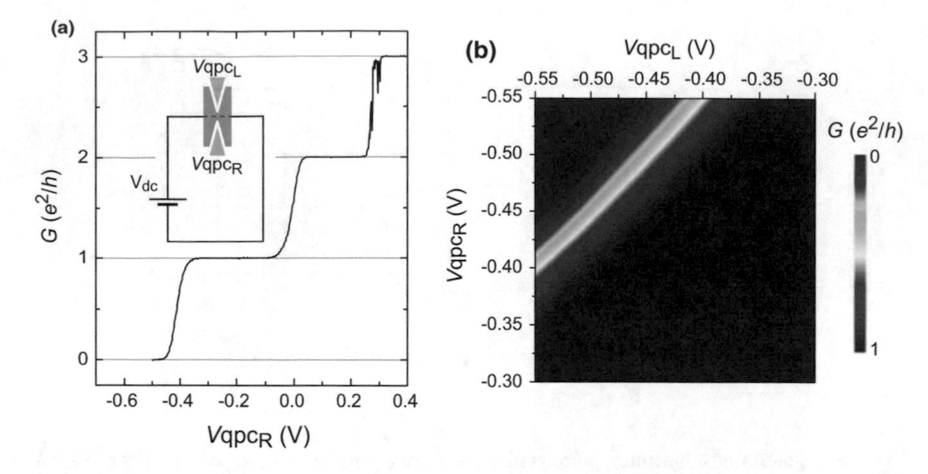

Fig. 1.3 Measurement of the conductance of a Quantum point contact. The intrinsic conductance of a QPC is measured with respect to the voltage applied on its two gates (left and right), in presence of a strong magnetic field (see text). **a** The conductance of a QPC is measured as a function of the right gate, the left gate is fixed (to $V_{qpc_L} = -0.55\ V$ for this trace). **b** Both gates are swept separately. A smooth step from 0 (in violet) to G_K (in red) is observed

This conductance has been measured when the QPC was directly connected to a voltage source as drawn in Figs. 1.2 and 1.3. We will call this the *intrinsic conductance*[4] of the QPC.

Note that the quantum of conductance $G_K = e^2/h \approx 1/25.8\,\mathrm{k}\Omega$ is a universal value. The practical way to build the quantum conductor does not matter: its chemical composition, the density of the 2DEG, the geometry of the split gates, the voltage applied on the QPC gates, etc. Other examples of quantum coherent conductors include e.g. atomic-size contacts [14] or carbon nanotubes [15].

1.2 Quantum Transport of Electricity

In this section, we introduce some useful concepts on the quantum transport of electricity. We will discuss successively quantum Hall effect, dynamical Coulomb blockade and quantum shot noise. The theory will be described briefly and will be illustrated with measurements. The experimental data shown in the two last subsections have been used to establish primary thermometers at $T \approx 6\,\mathrm{mK}$ (see our open-access publication [16] where we have measured the lowest electronic temperature in the world for a mesoscopic device of micrometer or nanometer scale with three different techniques).

[4]In presence of other components, the conductance of a coherent conductor is modified by the interactions with the environment.

1.2.1 Quantum Hall Effect

The quantum Hall effect occurs when a 2DEG is subject to a strong perpendicular magnetic field [2]. The conductance of such a system acquires a universal value given by quantum of conductance $G_K \triangleq e^2/h$ times a filling factor ν that is either an integer or a fraction: $G = \nu\, G_K$. As explained by Laughlin in [17], this effect is an emergent phenomenon which cannot be deduced from first principles. In particular, despite a clean sample is required, the defects play a crucial role since a sample which is invariant by translation cannot display quantum Hall effect [17]. The ground state of a fractional quantum Hall effect cannot be derived from a non-interacting electrons approach. This can lead to very complicated physics, for instance, the origin of the even-denominator filling factor $\nu = 5/2$ remains an open question ([18] and references therein). In this thesis we consider only integer filling factors.

Edge states and filling factor

For the integer quantum Hall effect (IQHE), a semi-classical approach is relevant (for a derivation, see [19]). The energy of the electrons is quantized in Landau levels $E_L = (n + 1/2)\hbar\omega_C$, where $\omega_C = eB/m$ is the cyclotron frequency (with m the mass of the electron and B the magnetic field). At a given Fermi energy E_F, only the levels with $E_L(n, B) \leq E_F$ will be filled and will participate to the electrical transport. This yields the quantum Hall plateaus of conductance versus the magnetic field shown in Fig. 1.4a. Each mode will carry a quantum of conductance G_K and will be located at the edge of the sample (see Fig. 1.4b).

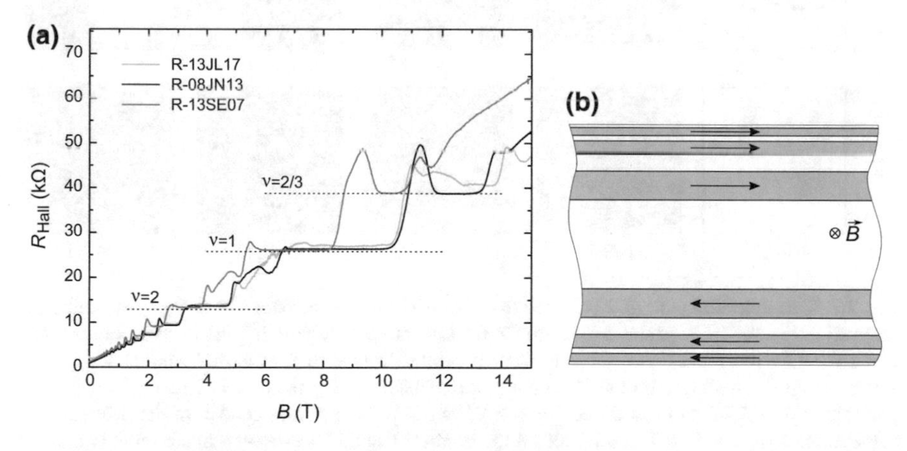

Fig. 1.4 Quantum Hall effect. **a** The Hall resistivity is plotted versus the perpendicular magnetic field for several 2DEGs (three traces). The horizontal straight dotted lines indicate the resistivity $R_{\text{Hall}} = \nu\, e^2/h$ expected for some filling factors. The mismatch comes from the resistance of the probes (a four-point probe method would be more accurate). The 2DEG used in this thesis is the '08JN13' (black trace). **b** Representation of the chiral edge states for a $\nu = 3$ integer filling factor

Percolation

The "bulk" of the 2DEG is insulating, the current flows only along the chiral edge states. The width and the distance to edge of these edge states depend on the magnetic field [20]. When passing from a filling factor to the lower one, the innermost channel percolates as illustrated in Fig. 1.5. This figure also shows that the screening of defects is better at higher filling factor, and, for a given filling factor, at the beginning of the quantum Hall plateau (low magnetic field). This can be observed when plotting the conductance of a QPC versus the voltage applied on its gates (as in Fig. 1.3) for different magnetic fields. In practice, we determine the best operating conditions to perform our measurements with canonical QPCs.

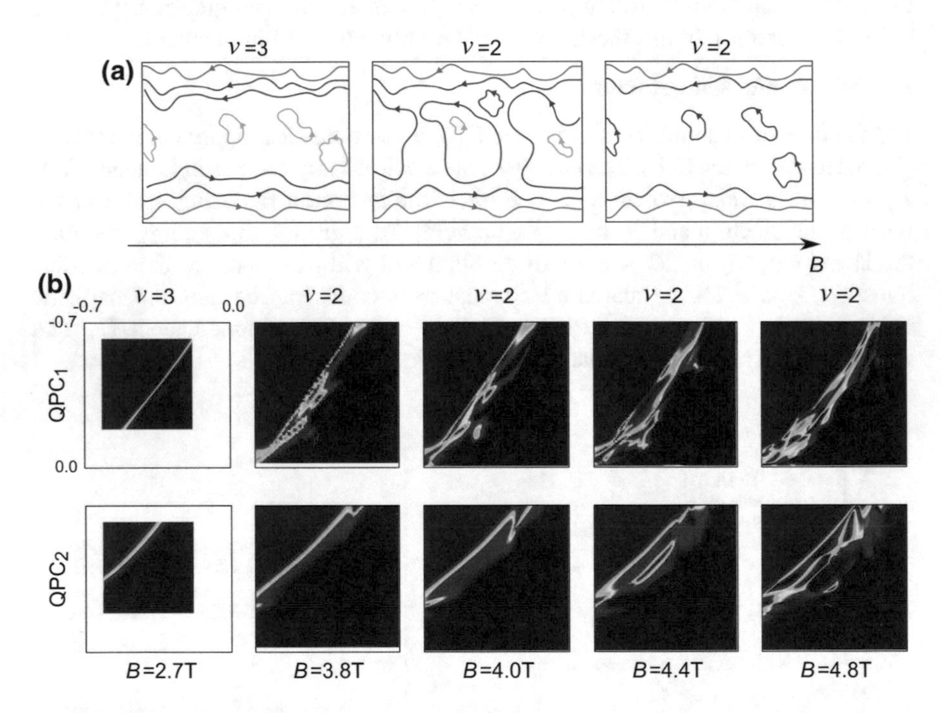

Fig. 1.5 Quantum Hall plateau. **a** (adapted from [21]) The filling factor can be changed by tuning the magnetic field B. The edge states percolate in the intermediate region. **b** The intrinsic conductance of QPC_1 (first row) and QPC_2 (second row) is plotted versus their two gate voltage $V_{qpc_{R,L}}$ for several magnetic field (see bottom line) but at the constant filling factor $\nu = 2$ (except for the left plot where $B = 2.7\,T$ and $\nu = 3$, see Fig. 1.4a). We use the same color code as in Fig. 1.3b and the same range $V_{qpc_{R,L}} \in [-0.7\,V, 0\,V]$ for all the graphs. The QPCs present a much more canonical shape in the beginning of the quantum Hall plateau (probably because the nearby defects are better screened by innermost channel)

1.2.2 Dynamical Coulomb Blockade (DCB)

Dynamical Coulomb blockade modifies the conductance of a quantum conductor at low energy when it is embedded in a true circuit (and not directly connected to a voltage source, as in Fig. 1.3). In general, the in situ conductance G of a quantum conductor will be reduced compared to its intrinsic value $G_\infty = \tau\, G_K \geqslant G$ because the electrons that go through the conductor will dissipate some energy into the circuit [22]. This phenomenon is closely linked to the granularity of the charge that tunnels through the quantum conductor, and it will disappear at integer transmissions $\tau = 1, 2 \ldots$ [23].

The problem can be solved in the case of a short tunnel junction in series with any impedance $Z(\omega)$ (see the same review [22]). In practice, a short tunnel junction is always a parallel combination of a capacitor and pure tunnel element, as sketched in Fig. 1.6. The geometry of the sample will set the capacitance C and thus the charging energy $E_C = e^2/(2C)$.

When dealing with resistive environments $Z(\omega) = R$, a charge that crosses the coherent conductor interacts with the electromagnetic modes of the environment. This interaction costs an energy that will change the transport properties of the coherent conductor at small voltage bias V_{dc} and low temperature. Indeed, in these conditions, the charge cannot pay the energy required to pass, it is then 'Coulomb blocked' and the differential conductance of the coherent conductor shows a dip when plotted versus the voltage bias V_{dc} (see Fig. 1.7c).

Theoretical predictions for a tunnel junction in series with a resistor

At zero temperature, a perturbative theory gives the flowing expression for the conductance of a small and opaque tunnel junction in series with an arbitrary resistor R at zero-temperature and with $e|V| \ll E_C$ (Eq. 113 in [22]):

$$G(V) = \frac{G_\infty (\pi/\gamma)^{2R/R_K} (1 + 2R/R_K)}{\Gamma(2 + 2R/R_K)} \left(R/R_K \frac{e|V|}{E_C} \right)^{2R/R_K} \tag{1.4}$$

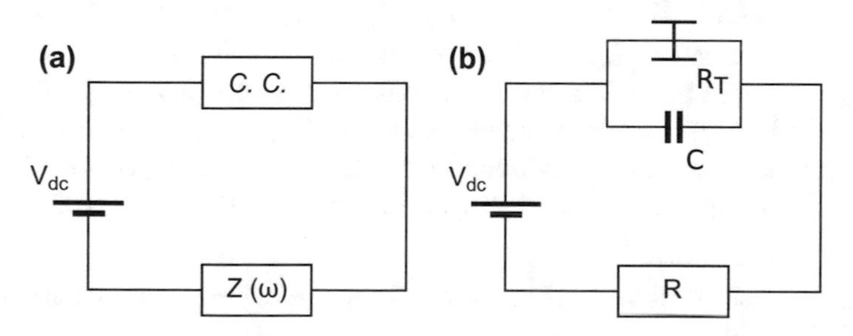

Fig. 1.6 A quantum coherent conductor in a circuit. **a** The general problem of a coherent conductor ($C.C.$) in series with a impedance $Z(\omega)$. **b** The theory is well-established in the case of low transmissions $\tau \ll 1$. Indeed the conductor can be separated in a pure tunnel element (of resistance R_T) in parallel with a capacitor C. We consider the case of a purely resistive impedance $Z(\omega) = R$

where $G_\infty \triangleq 1/R_T$ is the intrinsic conductance (as expected, we obtain $G = G_\infty$ for $R = 0$), Γ is the gamma function, $\gamma \approx \exp(0.5772)$ is the exponential of the Euler's constant.

There is also an expression of the conductance of the junction at zero bias and low temperature $k_B T \ll E_C$ [24]:

$$G(T) = \frac{G_\infty \pi^{(1/2+3R/R_K)} \Gamma(1 + R/R_K)}{2\Gamma(3/2 + R/R_K)} \left(R/R_K \frac{k_B T}{E_C} \right)^{2R/R_K} \tag{1.5}$$

The resistance R always appears in a ratio R/R_K. The strength of DCB depends on this ratio: (i) when $R \ll R_K$, $G(V) \sim G_\infty$, there is no Coulomb blockade (ii) when $R \gg R_K$, we are in the *static* Coulomb blockade regime,[5] the current is completely blocked unless the electrons have the charging energy $e|V| > E_C$ (iii) the ideal case to observe a fully developed DCB is when $R \sim R_K$.

From the two previous expressions, we can extract the experimental (absolute) temperature of a DCB measurement in the tunnel regime in the case (iii) with $R = R_K/2$. This measurement can be used to perform a *primary*[6] thermometry.

Experimental DCB measurement

We repeat the measurement twice, once with QPC_1 and once with QPC_2. To implement the circuit shown in Fig. 1.6b, we use QPC_i as a tunnel junction, and we set the opposite characterization gate to $\tau_{sw_j} = 2$. This switch will behave as a classical resistor[7] with a resistance $R = R_K/2$.

At low temperature, we measure the conductance versus the voltage bias V_{dc}. The conductance shows a dip at zero bias (see Fig. 1.7). The deeper the dip, the lower the temperature. At $R = R_K/2$, the expressions Eqs. (1.4) and (1.5) of the conductance are linear both in V (see the red dashed lines in Fig. 1.7c) and T:

$$\begin{cases} G(V) = \alpha \frac{G_\infty}{E_C/k_B} \times V \\ G(T) = \beta \frac{G_\infty}{E_C/k_B} \times T \end{cases} \tag{1.6}$$

where $\alpha = \frac{\pi e}{2\gamma}/k_B \approx 10.23 \times 10^3$ and $\beta = \frac{\pi^{5/2}}{8} \approx 2.18$ are constants. To be more accurate,[8] we can extract $E_C \approx 25\,\mu\text{eV} \approx 290\,\text{mK}$ from the height of the Coulomb diamonds shown in Fig. 1.7. The explanations on Coulomb diamonds will be given in Appendix B.2. From this numerical value of E_C, we can confirm that these DCB experiments were actually done in the tunnel regime: $G_{1\infty} \approx 0.12\, G_K$ and $G_{2\infty} \approx 0.13\, G_K \ll G_K$. Then we use a finite temperature theory of the tunnel

[5] An implementation of this limit will be discussed in Appendix B, where we consider the single electron transistor (SET).

[6] Such a thermometer does not require any calibration with a reference thermometer.

[7] An experiment reported in this thesis (in Fig. 4.10) shows that this is not true on the first plateau $R = R_K$.

[8] The temperature extracted will remain a primary thermometry.

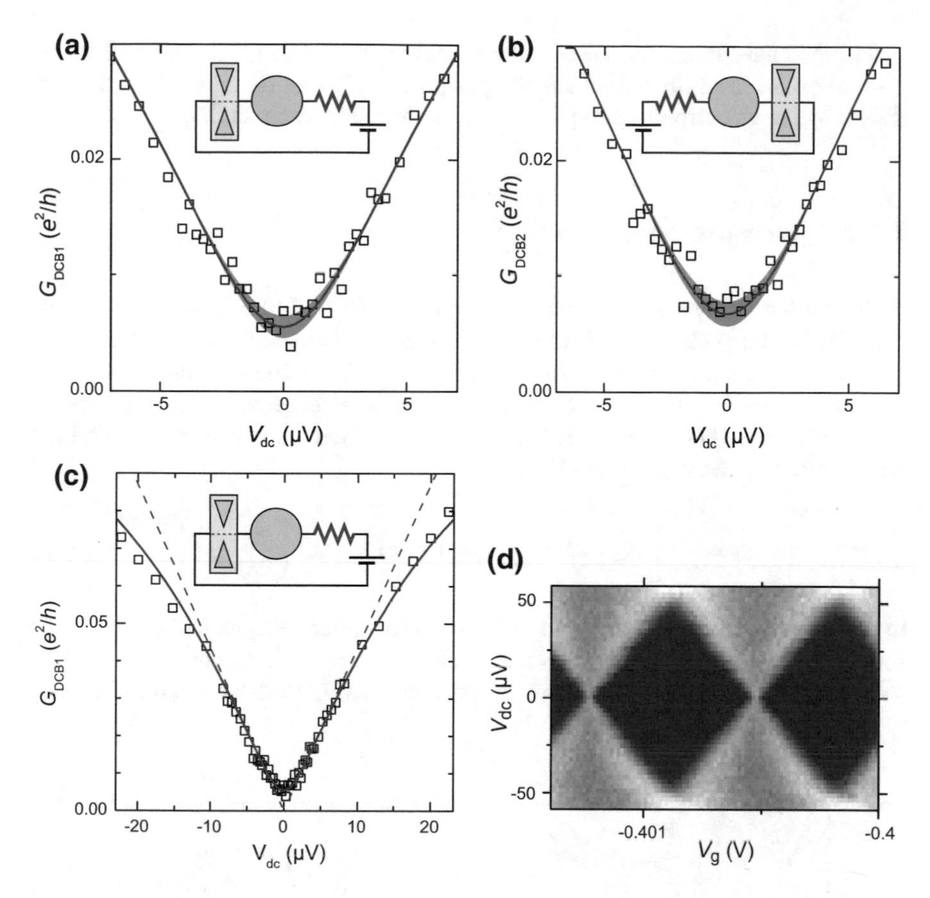

Fig. 1.7 Dynamical Coulomb blockade measurement. This measurement is done at $T \approx 6\,\text{mK}$ in the tunnel regime, with a resistive environment $R = R_K/2$. **a** and **b** With a resistor in series, the conductance G_{DCB} of a QPC is reduced at zero bias V_{dc}. The red line corresponds to the finite temperature theory for a tunnel junction at $T_{\text{DCB1}} = (6.0 \pm 1.0)$ mK for QPC_1 and $T_{\text{DCB2}} = (6.5 \pm 1.0)$ mK for QPC_2 (the grey shade shows the uncertainty ± 1 mK) **c** Same as **a** but displayed on a wider range in V_{dc}. **d** Coulomb diamond, its height gives the charging energy $E_C \approx 25\,\mu\text{V}$ (explanations given later, in Appendix B.2)

regime proposed by [25]. Their calculations have been plotted (as red solid lines) in Fig. 1.7a,b and c, where we have used $E_C = 25\,\mu\text{eV}$ and $T_{\text{DCB1}} = (6.0 \pm 1.0)$ mK for QPC_1 or $T_{\text{DCB2}} = (6.5 \pm 1.0)$ mK for QPC_2 to fit the data using G_∞ as a fitting parameter, the gray shade shows the uncertainty ± 1 mK.

We have seen that we are able to measure the temperature of a dynamical Coulomb blockade measurement in the tunnel regime with an environment $R = R_K/2$. This phenomenon is due to the granularity of the charge that flows through the tunnel junction made by a QPC. This granularity can be measured in the fluctuations of the current across the QPC. We will discuss this point in the next section.

Here we have discussed DCB in the tunnel regime using QPCs. This effect can also be explored in the strong coupling regime [26] or using atomic junction has shown by the Quantronics group (see [27], or the thesis of Cron [28]).

1.2.3 Quantum Shot Noise

So far, we have only discussed the average current. But, additional informations are contained in the fluctuations of current. We are going to measure these fluctuations to get a temperature measurement. In practice, we are able to measure the power spectrum of the current fluctuations S_I at zero frequency. This quantity can be computed using a scattering approach for a short coherent conductor connecting two terminals with a transmission τ (Eq. 62 in [29]):

$$S_I = \frac{e^2}{h}\left[4k_B T\tau^2 + 2eV_{\rm dc}\tau(1-\tau)\times\coth\left(\frac{eV_{\rm dc}}{2k_B T}\right)\right] \tag{1.7}$$

in this expression, the transmission $\tau(E) \approx \tau$ is assumed independent of the energy for $E \lesssim e|V_{\rm dc}|, k_B T$. The noise is maximal when $\tau = 1/2$. At small voltage bias $eV_{\rm dc} \ll k_B T$, the noise corresponds to the thermal (Nyquist-Johnson) noise for a

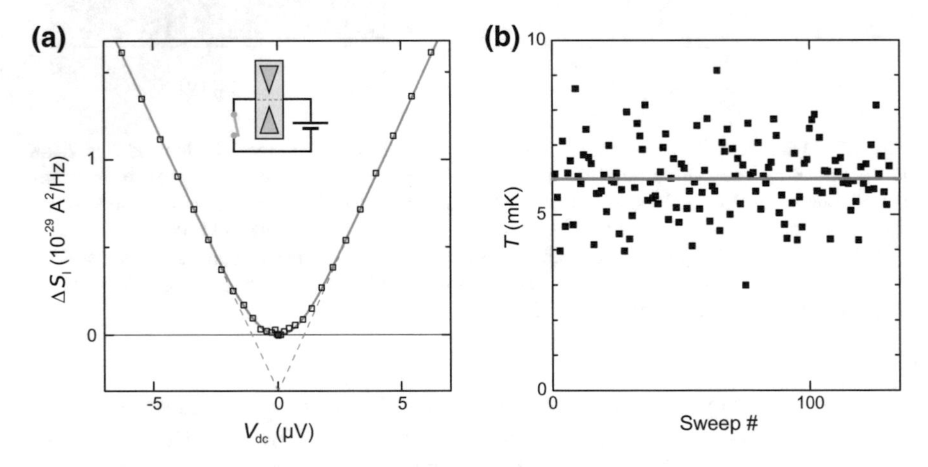

Fig. 1.8 Primary thermometry using quantum shot noise. **a** Power spectral density of current fluctuations in excess $\Delta S_I \triangleq S_I - S_I(V_{\rm dc} = 0)$ versus voltage bias $V_{\rm dc}$, for a quantum conductor at $\tau \approx 0.55$ directly connected to a voltage source (see the inset), and averaged on 131 sweeps. Solid gray line shows a fit of the averaged data. We obtain $T = (6.0 \pm 0.1)$ mK. The dashed lines correspond to zero temperature. **b** Temperature obtained to fit the data of each successive sweep of $V_{\rm dc}$. The solid line indicates the average

classical resistor $S_I = 4k_B T G_\infty$ where $G_\infty \triangleq \tau e^2/h$ is the intrinsic[9] conductance of the quantum conductor. This is also the case when $\tau = 1$, since the second term vanishes and $G_\infty = e^2/h$.

In Fig. 1.8, we set $\tau \approx 0.5$ and we use Eq. (1.7) to extract the temperature T. This is a primary thermometry as we are able to characterize the other parameters (V_{dc}, τ). The temperature $T = (6.0 \pm 0.1)$ mK is essentially given by the curvature of the excess noise ΔS_I at low voltage bias. The accuracy on the measurement can be significantly enhanced by statistically averaging the measure on many sweeps (or, in principle equivalently, by increasing the integration time). The precision is then limited by the stability of the experimental setup on long time scales (the Fig. 1.8a requires ~45 h of acquisition).

1.3 Summary of This Thesis

This thesis reports electrical transport measurements of a tunable and characterizable nano-device (shown in Fig. 1.9). Owing to its hybrid metal-semiconductor structure and to multiple gates, this device provides a quantitative testbed to strongly correlated and critical physics. It can be tuned to a regime where a few independent electronic channels are strongly interacting with the 'charge' macroscopic quantum degrees of freedom of the metallic node (in purple in Fig. 1.9) of the circuit. We used this sample to probe and to control the degree of charge quantization on a metallic node, depending on its connection to other conductors. The device can also be tuned to implement the multi-channel Kondo model and perform reliable quantum simulations of this quantum many-body model. As such, it provides an experimental testbed for some of the most powerful quantum many-body techniques (the numerical renormalization group [30–32], Bethe ansatz [19, 33], conformal field theory [35–37] or bosonization [38–40]).

We have observed good agreements between our experimental data and the predicted universal power laws both near the strong coupling limit in the problem of the charge quantization and in the vicinity of the quantum critical points of the multi-channel Kondo effect. The crossover from these critical points reveals intriguing physics, in particular at intermediate temperature. The approach developed in this thesis paves the way to further study of the tantalizing non-Fermi liquids physics underlying the field of strongly correlated materials.

The remainder of this introductory section gives a quick preview of the results reported in this thesis, with each subsection corresponding to a chapter. We start with a description of our single-electron device and answer a long standing question on the criterion of how the charge quantization is destroyed by the quantum fluctuations in such a device. In the next section, we will see that this system can be mapped onto the Kondo model using the 'charge' degrees of freedom of the circuit. Under

[9]In this experiment, the quantum conductor is directly connected to a voltage source (see the inset of Fig. 1.8a).

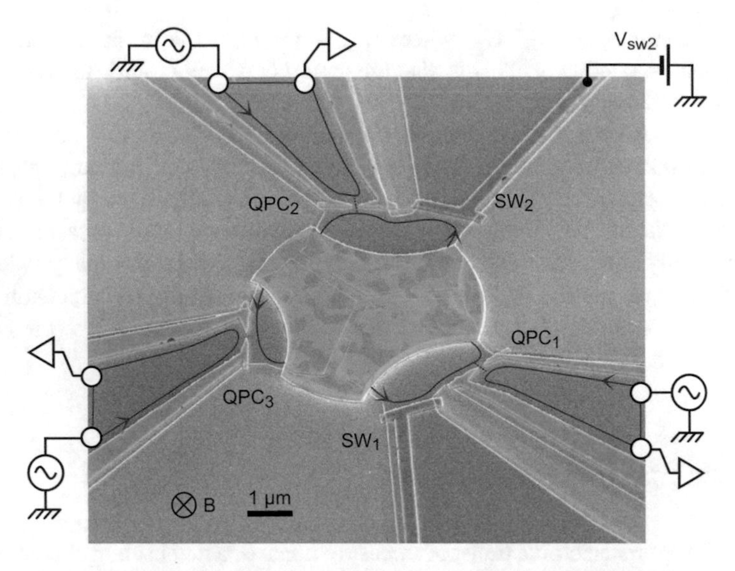

Fig. 1.9 Colored micrograph of the sample with measurement schematics. This figure shows the central ohmic contact (in purple) that redistributes the current injected by the a.c. voltage sources (out of the image) into the chiral edge states of the IQHE (red lines) through larger ohmic contacts (white circles) not shown in the picture. The low frequency signals are measured using Lock-in amplifiers (triangles). This sample includes three QPCs (in cyan) and two switch gates (in orange) used for characterization. The value of the transmission of the QPC and the state of the switch is controlled by the voltage sources that connect (black circles) theses surface gates

the renormalization process, its conductance is predicted to flow towards non-trivial Kondo fixed points at low temperature. In the next and last section, we focus on the quantum critical scalings that occur when we purposely introduce perturbations that are 'relevant' in the renormalization group sense.

1.3.1 Control of the Charge Quantization

A single-electron transistor in the strong coupling limit

The central piece of metal shown in purple in Fig. 1.9 is the main character of this thesis. It will be called the *island*, since it is connected to the circuit by only a few electronic channels. The geometry of the sample sets the typical charging energy $E_C \triangleq e^2/(2C)$ (with e the elementary charge and C the geometrical capacitance of the island) required to add/remove an electron to/from the island. Single-electron effects are thus important at temperatures and energies below this scale: $k_B T, eV \ll E_C \approx 25\,\mu\text{eV} \approx k_B \times 300\,\text{mK}$.

The connection between the island and the electronic circuit (constituted of large electronic reservoirs) can be adjusted through tunable QPCs (in cyan). Here we

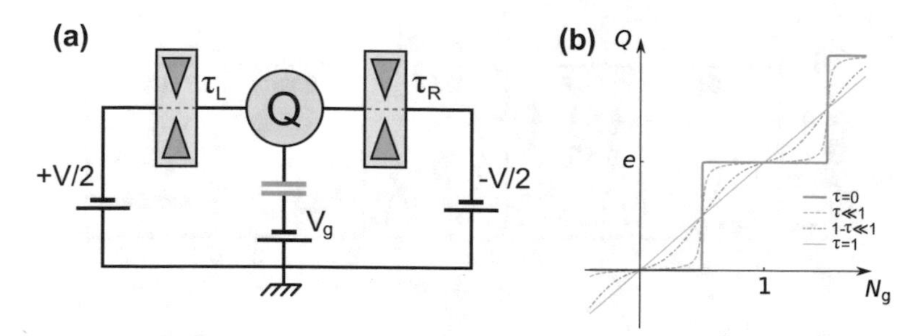

Fig. 1.10 Charge quantization in a single-electron device. **a** Schematics of the sample. The differential conductance G_{SET} is measured by applying an a.c. voltage bias to the device. In presence of charge quantization, this conductance will show Coulomb oscillations with respect to a capacitively coupled gate V_g. **b** (adapted from [39]) The average charge Q on the island is plotted at zero-temperature versus $N_g \triangleq V_g/\Delta V_g$ (where ΔV_g is the period of the Coulomb oscillations) for different connection strength τ (the transmission probability $\tau \triangleq \tau_R$, with $\tau_L = 0$). The charge quantization completely disappears as soon as there is one ballistic electronic channel connected to the island ($\tau = 1$)

consider only two QPCs (see Fig. 1.10a). When the transmissions of both left and right QPCs are set to the tunnel regime $\tau_{L,R} \ll 1$, the island is only weakly connected to the surroundings and its charge is quantized in units of e. In this limit, the device implements the well-known single-electron transistor (SET). The number of charges on the island can be tuned separately by sweeping the voltage V_g of a lateral gate (e.g. a characterization gate, in orange in Fig. 1.9).

One would then observe Coulomb blockade oscillations of the conductance of the device $G_{\text{SET}}(V_g)$ as current is allowed only when two successive charge states of the island are degenerate in energy (up to thermal fluctuations, otherwise the charge state is frozen and the current is blocked). The number of electrons on the island is thus incremented by one after each peak of conductance shown in Fig. 1.11a. In this figure, we see that the oscillations progressively vanish as the transmission of a QPC tends to unity $\tau \longrightarrow 1$. A more systematic study consists in plotting the visibility of the oscillations $\Delta Q \triangleq (G_{\text{SET}}^{\max} - G_{\text{SET}}^{\min})/(G_{\text{SET}}^{\max} + G_{\text{SET}}^{\min})$ as a function of τ_R. All of the traces obtained for different values of a fixed τ_L collapse when the transmission becomes ballistic at $\tau = 1$, and no Coulomb oscillation is observed above this limit (see Fig. 1.11b).

It is known theoretically that quantum fluctuations of the charge smear the charge quantization (even at zero temperature) [41]. In 1995, Matveev showed that quantization is completely destroyed in the ballistic limit when considering an island with a continuous density of states (see Fig. 1.10b) [39]. In 1993, a controversy arose regarding this criterion for the destruction of charge quantization, as some experimentalists validated it [42] whereas others observed Coulomb oscillations above the ballistic limit [43]. But these early experiments were based on non-metallic islands with discrete density of states, and where the phase coherence of the electrons could give rise to subtle mesoscopic effects [44]. The design and the materials we used to

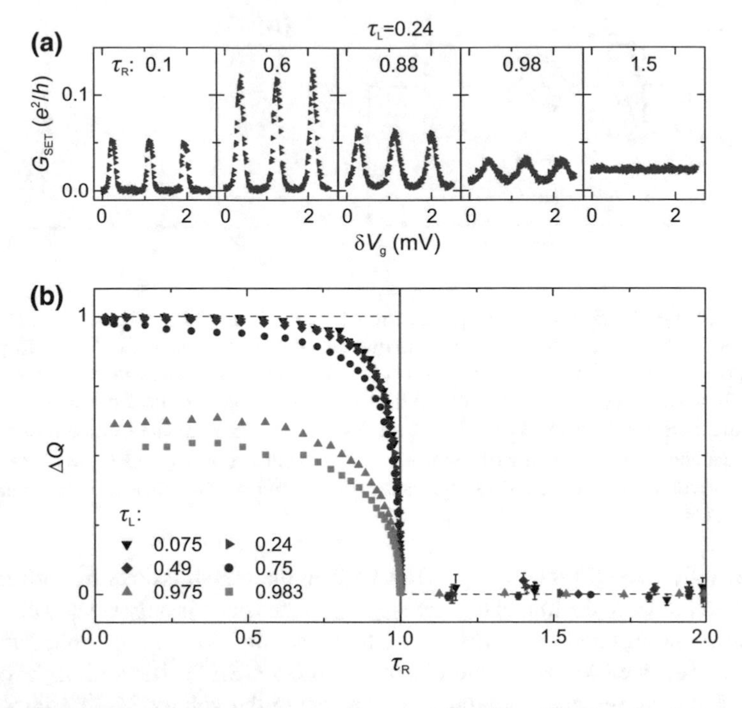

Fig. 1.11 Charge quantization versus connection strength at $T \simeq 17$ mK. **a** Conductance sweeps $G_{SET}(\delta V_g)$ with a fixed $\tau_L = 0.24$, and $\tau_R = 0.1, 0.6, 0.88, 0.98$ and 1.5, from left to right respectively. **b** Visibility of G_{SET} oscillations versus τ_R, with each set of symbols corresponding to a different QPC$_L$ set-point

fabricate our sample avoid any such coherent effects, since when an electron enters the island, it stays there for much longer than its quantum phase coherence time.

Quantitative comparison with theory

Assuming a continuous density of state in the island and spinless[10] electrons, quantitative predictions can be established for this system in several limits. At low temperature $k_B T \ll E_C$, we can compare the visibility ΔQ of the conductance oscillations when approaching the ballistic limit $\tau_R \longrightarrow 1$ with the quantitative theory in two limits (strong coupling $1 - \tau_L \ll 1$ [40] and asymmetric $\tau_L \ll 1$ [45]). We are able to characterize all the parameters (τ_L, τ_R, E_C and T) independently, and the quantitative agreement found in the strong coupling limit is therefore established without any fitting parameters (see Fig. 1.12a).

In the two limits, the theory predicts a $\sqrt{1 - \tau_R}$ dependence for the visibility of the Coulomb oscillations. This power law behavior has been observed also for intermediate transmissions of $\tau_L \in [0, 1]$. At higher temperatures, this dependence becomes completely universal on the full range $\tau_{R,L} \in [0, 1]$ (as indicated by the

[10]The spin degeneracy is lifted with a strong magnetic field.

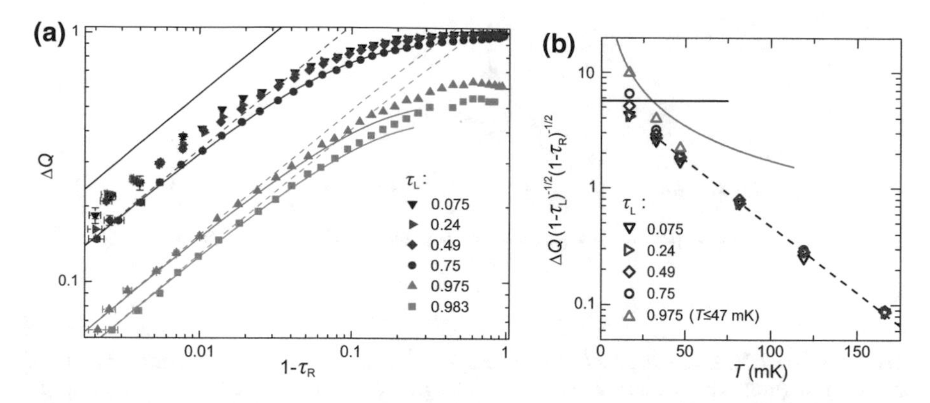

Fig. 1.12 Charge quantization scaling near the ballistic point and its exponential suppression with temperature. **a** The visibility ΔQ of the Coulomb oscillations at $T \approx 17\,\mathrm{mK}$ is displayed versus $1-\tau_R$ in a log-log scale, with distinct sets of symbols for the different QPC_L set-points. Solid lines are quantitative predictions (no fit parameters) derived assuming $k_B T \ll E_C$, $1-\tau_R \ll 1$ and either $\tau_L \ll 1$ (top straight line) or $1-\tau_L \ll 1$ (three bottom continuous lines). The power law $\Delta Q \propto \sqrt{1-\tau_R}$ (straight dashed lines) is systematically observed for $1-\tau_R \lesssim 0.02$, also at intermediate τ_L. **b** symbols display versus T, in semilog scale, the rescaled data $\Delta Q/\sqrt{(1-\tau_R)(1-\tau_L)}$, extracted in the regime of small enough $1-\tau_R$ such that $\Delta Q \propto \sqrt{1-\tau_R}$. Solid lines correspond to the quantitative predictions in the quantum regime $k_B T \ll E_C$ shown in **a**. The straight dashed line displays an exponential decay close to predictions in the presence of strong thermal fluctuations

collapse of all the τ_L data points in Fig. 1.12b). Moreover, in Fig. 1.12b, we note that the charge quantization (measured by ΔQ) is also exponentially suppressed with the temperature T, as expected by the theory in presence of strong thermal fluctuations $k_B T \gg E_C/\pi^2$ ([46] and references therein).

1.3.2 Observation of the Multi-channel Kondo Effect

The original Kondo model and some of its variations

In 1964, Kondo computed the contribution to the resistivity of the scattering of conduction electrons by magnetic impurities in dilute alloys [47]. However, his perturbative approach fails at low temperatures compared to the so-called Kondo temperature T_K. This is a typical problem suited for the renormalization group theory. Its first exact solution was found by Wilson using numerical renormalization group [30].

The original Kondo model is illustrated in Fig. 1.13. It describes a magnetic impurity (modeled by a spin $\vec{S}$, $S = 1/2$ throughout this thesis) that interacts with a single band of conduction electrons (shown as a 1D lattice) through an on-site local antiferromagnetic coupling J. At low temperature J is effectively renormalized to $J \longrightarrow \infty$ [30]. In this limit, the impurity forms a singlet with an electron, and this complete screening yields a simple Fermi liquid description [48]. Nozières and Blandin proposed a generalization of the model with N *independent* electronic channels [49],

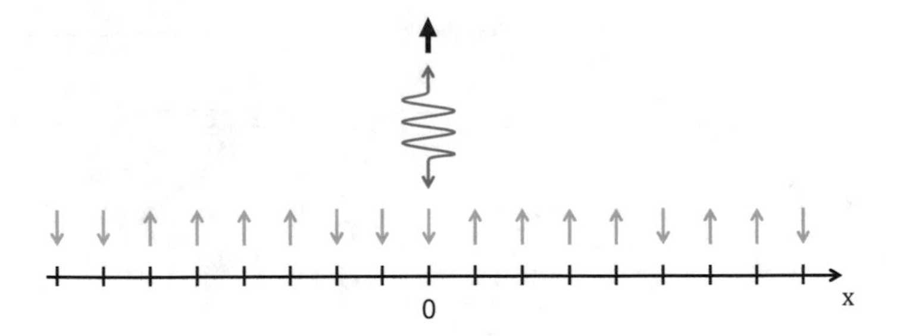

Fig. 1.13 Diagram explaining the Kondo model. The impurity $S = 1/2$ is represented by a thick black arrow (top). The conduction electrons (light gray arrows) are distributed on a lattice. To simplify, the lattice is 1D. The antiferromagnetic interaction is drawn as a squiggly arrow on the impurity site

which leads to non-Fermi liquid ground state at low temperature in the overscreened case $N > 2S$. In the renormalization group picture, the couplings J_i between the quantum impurity and each electronic channel are effectively renormalized as the temperature is lowered [50]. At zero-temperature, they eventually reach a universal Kondo fixed point, which depends on the number of channels N [49].

In our device, two successive charge states can be tuned (using V_g) to have the same energy. The number of electrons on the island n or $n + 1$ can then play the role of a quantum two-level system, or, in other words, of a pseudo-spin $S = 1/2$. In 1991, Matveev demonstrated, in the weak coupling limit, the exact mapping between the original 'spin' Kondo model and the Coulomb blockade that models our system [51]. This mapping, which is also valid beyond the tunnel limit [39, 52], involves the coupling between the Kondo 'charge' pseudo-spin and a localization pseudo-spin of the electrons (either in or out of the island).

Observation of multi-channel Kondo effects and universal behaviors

The conductance G_i of each QPC will be renormalized towards the 'charge' Kondo fixed points at low temperature. In Fig. 1.14, when we tune all the transmissions to be in use equal ($\tau \triangleq \tau_1 = \tau_3$ and $\tau_2 = 0$ for two-channel (2CK); $\tau \triangleq \tau_1 = \tau_2 = \tau_3$ for three-channel (3CK)) and the gate voltage to charge degeneracy ($\delta V_g = 0$), we observe that the conductance flows to the predicted 'charge' Kondo fixed point (which is extremal $G^*_{2CK} = e^2/h$ for two-channel and intermediate $G^*_{3CK} = 2\sin^2(\pi/5)e^2/h \approx 0.691e^2/h$ for three-channel) [53]. Note that these fixed points are universal, i.e. they do not depend on any microscopic parameters (transmission τ, charging energy E_C, etc.).

The scaling and universal properties are inherent to renormalization [30]. In the Kondo model, the temperature evolution of any observable is a universal function of the rescaled temperature T/T_K, provided the temperature has been lowered enough so that the renormalization has suppressed the influence of the 'irrelevant' perturbations [54]. Mitchell and co-workers have computed the full universal curve of

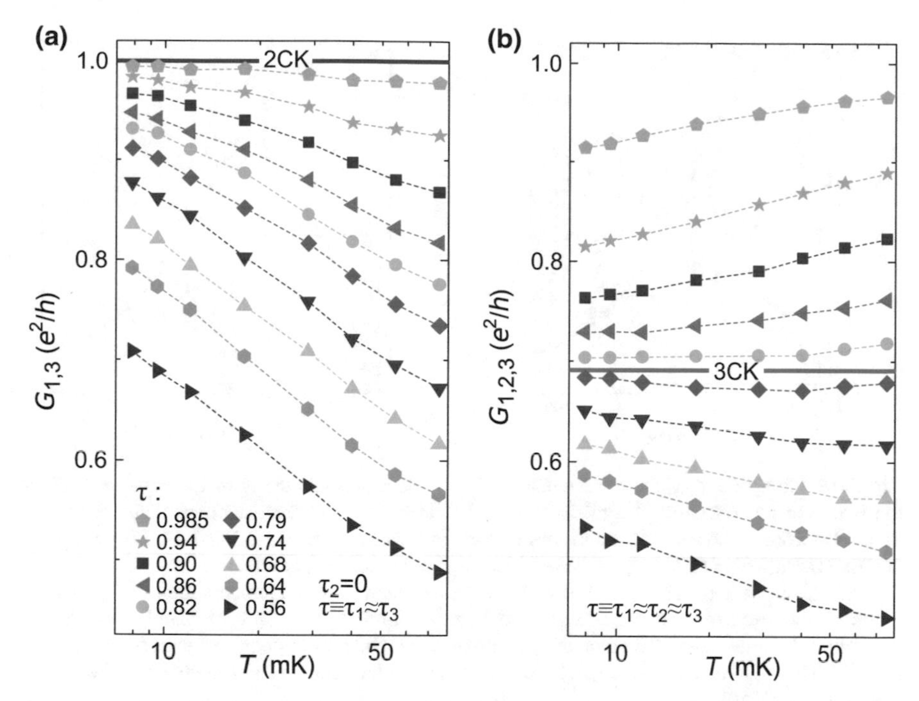

Fig. 1.14 Renormalization towards the two- and three-channel 'charge' Kondo fixed point. For the same set of transmissions τ, the individual conductances at degeneracy $\delta V_g = 0$ are plotted versus the temperature $T \approx \{7.9, 9.5, 12, 18, 29, 40, 55, 75\}$ mK in log-scale. **a** when lowering the temperature, the conductance flows towards the 2CK fixed point (red thick line) for all the transmissions in the case of two symmetric channels ($\tau_1 \approx \tau_3$). **b** with three symmetric channels, it flows to G^*_{3CK} (green thick line)

conductance $G(T/T_K)$ (from $G(T/T_K \gg 1) \approx 0$ to $G(T/T_K \ll 1) \approx G^*$) for both the two- and three-channel 'charge' Kondo models [55, 56]. We compare our experimental data to their exact numerical calculations in Fig. 1.15. Our procedure is to rescale the temperature to match the point at base temperature with the theoretical curve. We note that at least the three lowest temperature points are in the universal regime as they follow the theoretical curve. This procedure works on many orders of magnitude in T/T_K, highlighting the outstanding tunability of our sample. At higher temperatures, non-universal effects appear. In fact, this crossover to a non-universal regime is fully explained by numerical renormalization group calculations taking into account the finite charging energy E_C. From this rescaling in T/T_K, one can also extract the relation between the Kondo temperature T_K and the unrenormalized coupling strength τ. In the insets of Fig. 1.15, we compare T_K to the theories near the fixed points and in the tunnel regime.

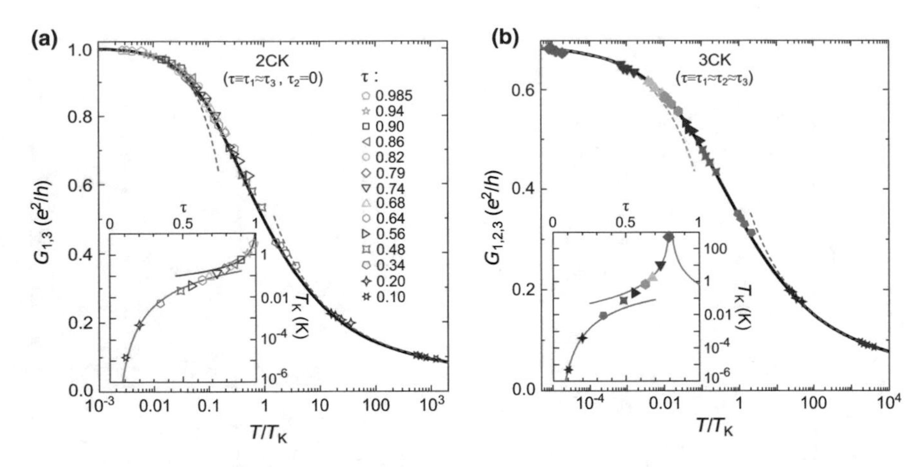

Fig. 1.15 Comparison of the experimental data to the universal curves of the conductance for the two- and two-channel 'charge' Kondo model. Here, experimental data are displayed up to $T = 29\,\mathrm{mK}$ and some additional transmissions are shown as compared to Fig. 1.14. Each set point at τ fixed is shifted horizontally in the semi-log representation so that the lowest temperature point matches the theoretical curve (solid black line). This defines a scaling temperatures $T_K(\tau)$ that are plotted in the insets for both the 2CK (**a**) and 3CK (**b**) configurations and compared to theoretical predictions (colored lines). The blue lines in the tunnel regime correspond to a perturbative theory [40, 50]. The colored dashed lines are the predicted power laws for the conductance near the Kondo fixed points [40, 56]

1.3.3 Quantum Phase Transition in Multi-channel Kondo Systems

Quantum criticality

The work on quantum phase transitions has been mainly motivated by one of the most important unsolved problem in condensed matter physics, which concerns the complicated phase diagram of some strongly correlated materials. In particular, the 'strange metal' phase from which the superconductivity of high-T_C superconductors emerges at low temperature has attracted a lot of theoretical effort but remains poorly understood (see [57] for a recent review and Fig. 1.16a). This phase displays signatures of quantum criticality such as a non-Fermi liquid power law of the resistivity versus temperature and a widening of the critical interval with temperature.

In contrast to classical critical phenomena that occur at the critical temperature T_C of a second-order phase transition, quantum criticality originates from quantum fluctuations that exist even at zero temperature at a quantum critical point. A physical system is driven to quantum criticality thanks to a non-thermal parameter g (e.g. the doping as in Fig. 1.16a, the pressure, or a gate voltage). It will exhibit quantum criticality for a range of parameter that starts from $g = g_c$ at zero-temperature, and which widens as a power law $T_{\mathrm{co}} \propto (g - g_c)^\gamma$, where γ is called 'critical exponent' and

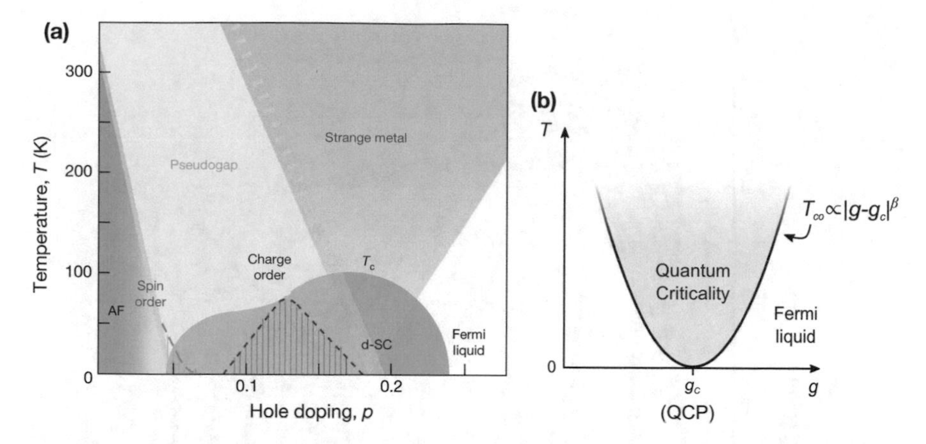

Fig. 1.16 Typical phase diagrams. **a** (reproduced and simplified from [57]) Typical phase diagram of a copper oxide plotted versus hole doping. At low temperature, depending on the hole doping, these systems display either an antiferromagnetic order (AF, in blue), a spin order (green strips), a charge order (red strips) or a d-wave superconducting order (d-SC, in green). A Fermi liquid is obtained in the overdoped regime at low temperature. **b** Typical phase diagram displaying quantum criticality when the non-thermal parameter g is tuned around the quantum critical point (QCP) at low temperature

T_{co} is the crossover temperature below which the system escapes quantum criticality (see Fig. 1.16b) [58, 59].

Despite the opportunity offered by tunable nano-devices for comparisons with theory, the realizations of quantum criticality in such systems are rare [60–62]. Exact predictions exist for the overscreened Kondo fixed points, which are known for their non-trivial critical exponents on various physical quantities ([36, 37] and references therein). In our 'charge' implementation, we can observe a crossover from quantum criticality by either introducing a channel asymmetry on the transmissions τ or by detuning from charge degeneracy $\delta V_g = 0$.

Development of a quantum phase transition

In Fig. 1.17, we set $\delta V_g = 0$, and we plot the conductance G versus temperature. Each arrow points towards low temperatures and corresponds to a given configuration (τ_2; $\tau_1 = \tau_3$). This graph therefore provides a visualization of the three-channel Kondo renormalization flow (with two channels set symmetric $\tau_1 = \tau_3$). Depending on the number of symmetric channels that share the largest bare connection τ with the island, the individual conductances G flow towards one of the one- (blue disk), two- (red disk) or three-channel (green disk) Kondo fixed points.

For symmetric transmission configurations $\tau_1 = \tau_2 = \tau_3$, along the diagonal, the individual conductances G remain symmetric at all temperatures. However, this setting is visibly critical since any initial asymmetry grows as the temperature is lowered. This graph therefore provides a direct view on the development of a quantum phase transition.

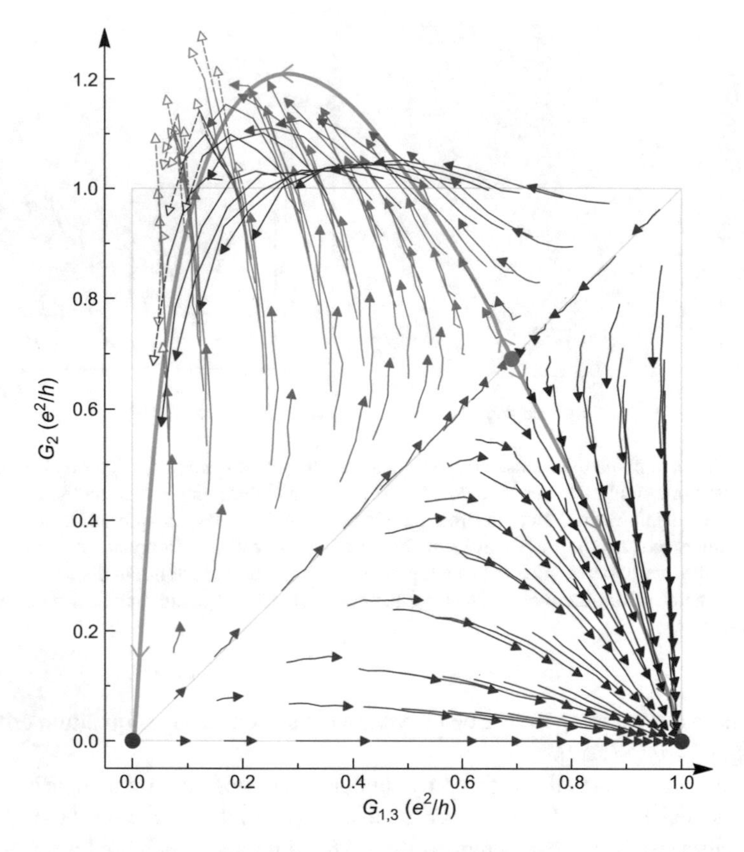

Fig. 1.17 Observation of the 3CK renormalization flow diagram. The averaged conductance of G_1 and G_3 is plotted versus G_2 at temperatures $T \approx \{7.9, 9.5, 12, 18, 29, 40, 55\}$ mK. The color of the arrows maps their orientation. The uncertainty on the open symbols and dashed lines is smaller than $0.1\, e^2/h$ whereas it is smaller than $0.05\, e^2/h$ for the solid symbols and lines. The Kondo fixed points are indicated with colored dots (1CK in cyan, 2CK in red and 3CK in green). The grey lines are NRG calculations

In the weak coupling regime $\tau_{1,2,3} \ll 1$, where all the channels can be treated independently and with a perturbative theory [50], an increase of the conductances is predicted [40]. The position of the one-channel fixed point at $J^*_{\rm 1CK} \longrightarrow \infty$ in the 'spin' Kondo effect corresponds unexpectedly to $G^*_{1CK} = 0$ for our device. This yields the non-monotonic behaviors and the arrow crossings visible in the upper part of the diagram ($\tau_2 > \tau_{1,3}$).

A remarkable overstepping of the quantum of conductance by the individual conductance of a single channel is also observed in the flow towards the one-channel fixed point. This observation is in agreement with the recent numerical renormalization group calculations of Mitchell [56]. Another important feature of this diagram is the visualization of the crossover from the 3CK non-Fermi liquid fixed point to the 2CK non-Fermi liquid fixed point.

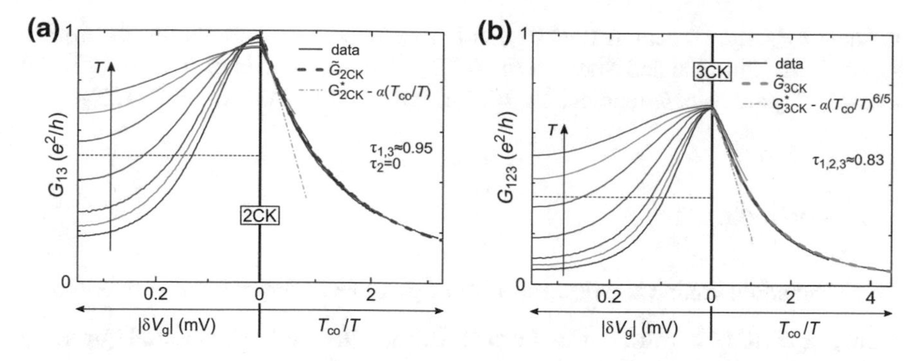

Fig. 1.18 Crossover from quantum criticality when detuning from charge degeneracy. A peak of conductance is shown at $T \approx \{7.9, 9.5, 12, 18, 29, 40, 55\}$ mK for a selected transmission τ close to the 2CK (in **a**) and the 3CK (in **b**) fixed points. The solid lines are experimental data. For each graph **a** and **b**, two scales are used: the raw plunger gate voltage δV_g (left) and a rescaled axis T_{co}/T, with $T_{co} \propto \delta V_g^{\beta}$ (right). The conductance adopts a universal behavior in the rescaled axis (except for the highest two temperatures). The $\tilde{G}_{2CK}$ red dashed line in **a** shows the zero temperature analytical prediction for $\tau = 0.95$. In **b**, $\tilde{G}_{3CK}$ is an NRG prediction (where the proportional constant in T_{co} is a fit parameter). Dashed-dotted grey lines are $(T_{co}/T)^{2/\beta}$ power-laws valid at $T_{co}/T \ll 1$

Crossing over from quantum criticality using an effective magnetic field

In this section, we set symmetric transmissions (either $\tau \triangleq \tau_1 = \tau_3$ and $\tau_2 = 0$ or $\tau \triangleq \tau_1 = \tau_2 = \tau_3$). The fragile non-Fermi liquid quantum critical state obtained at low temperature $T \ll T_K, E_C$ is destroyed by detuning the device from charge degeneracy. A non-zero δV_g would favor one state of the charge pseudo-spin state and destroy the Kondo effect as an effective magnetic field [51]. This would drive the system to a Fermi liquid state with a typical energy scale given by the crossover temperature T_{co} (see Fig. 1.16b). This quantity is predicted to depend on δV_g as a power law $T_{co} \propto \delta V_g^{\beta(N)}$ with a critical exponent $\beta(N) = (2 + N)/N$ [63].

Moreover, in the vicinity of the quantum critical point, T_{co} is the only energy scale to consider, and the conductance is expected to follow a universal function of T/T_{co} [55, 64]. The quantitative expression of $T_{co}(\Delta E)$ with respect to the level splitting[11] ΔE have been computed by Furusaki and Matveev for the 'charge' 2CK model for arbitrary ΔE (beyond the power law at $\Delta E \ll E_C$) [40]. In Fig. 1.18a, we use this expression to rescale δV_g and we observe that the conductance peaks for different temperatures all collapse on the same universal curve. Knowing the critical exponent $\beta(N)$, we have proposed a naive generalization of this rescaling that yields a collapse of our 3CK conductance peaks (see Fig. 1.15b) and which is in agreement with NRG calculations where we used the multiplicative constance in T_{co} as a fit parameter. We therefore verify the predicted universal power law for the crossover temperature T_{co}

[11]As explained above, δV_g acts as a magnetic field. We define $\Delta E \triangleq 2E_C\delta V_g/\Delta V_g$, where $E_C/k_B \approx 300$ mK is the charging energy and $\Delta V_g \approx 0.70$ mV is the period of the Coulomb blockade oscillations (these two numerical values are determined by the geometry of the sample and are independent of T or τ).

at small δV_g and we demonstrate a quantitative agreement with the full theoretical prediction of Furusaki and Matveev for 2CK (red dashed line in Fig. 1.18a) [40] and a single-fit-parameter agreement for 3CK in (green dashed line in Fig. 1.18b).

1.4 Outlook

This concluding section is written as a list of possible further research directions.

Charge quantization and multi-channel Kondo effect in the fractional quantum Hall effect regime

As shown in Fig. 1.4, the 2DEG we use displays a quantum Hall plateau at the fractional filling factor $\nu = 2/3$. The behaviors of the QPCs and the connection of the electronic channels to the central micron-size ohmic contact are probably different at this filling factor. However, it would be very interesting to measure the destruction of the charge quantization and the renormalization flow of the conductance towards the multi-channel Kondo fixed point in the FQHE regime.

Noise measurement near the quantum critical Kondo fixed points

Measuring the noise on the current can give access to the statistics of the charge carriers [65]. The non-Fermi liquid physics near the quantum critical points of the overscreened Kondo fixed point could lead to a fractionalization of the charge [66]. Probing this exotic effect should be possible with shot noise measurements. This noise vanishes in principle near the 2CK fixed point (which is located at ballistic transmission). However the 3CK fixed point should be accessible.

Two-impurity sample

The sample studied in this thesis counts a single pseudo-spin Kondo impurity. Nanofabricating a new sample with two metallic islands separated by a QPC may implement the two-impurity and the multi-channel two-impurity 'charge' Kondo model [67]. The two-impurity model is important as it includes both an RKKY interaction between the impurity spins and the usual Kondo interaction between each impurity spin and the conduction electrons [68]. This is a simplified version of the Kondo lattice which is central for heavy fermions [69].

References

1. M.H. Devoret, H. Grabert. *Single Charge Tunneling: Coulomb Blockade Phenomena in Nanostructures: [Proceedings of the NATO Advanced Study Institute on Single Charge Tunneling, held March 5–15, 1991, in Les Houches, France]* (Plenum Press, 1992)
2. K. von Klitzing, G. Dorda, M. Pepper, New method for high-accuracy determination of the fine-structure constant based on quantized hall resistance. Phys. Rev. Lett. **45**(6), 494 (1980)

3. R.A. Webb, S. Washburn, C.P. Umbach, R.B. Laibowitz, Observation of h/e Aharonov-Bohm oscillations in normal-metal rings. Phys. Rev. Lett. **54**(25), 2696 (1985)
4. Y.V. Nazarov, Y.M. Blanter, *Quantum Transport: Introduction to Nanoscience* (Cambridge University Press, Cambridge, 2009)
5. R. Landauer, Spatial variation of currents and fields due to localized scatterers in metallic conduction. IBM J. Res. Dev. **1**(3), 223–231 (1957)
6. M. Büttiker, Y. Imry, R. Landauer, S. Pinhas, Generalized many-channel conductance formula with application to small rings. Phys. Rev. B **31**(10), 6207 (1985)
7. M. Büttiker, Four-terminal phase-coherent conductance. Phys. Rev. Lett. **57**(14), 1761 (1986)
8. B.J. van Wees, H. van Houten, C.W.J. Beenakker, J. G. Williamson, L.P. Kouwenhoven, D. van der Marel, C.T. Foxon, Quantized conductance of point contacts in a two-dimensional electron gas. Phys. Rev. Lett. **60**(9), 848 (1988)
9. D.A. Wharam, T.J. Thornton, R. Newbury, M. Pepper, H. Ahmed, J.E.F. Frost, D.G. Hasko, D.C. Peacock, D.A. Ritchie, G.A.C. Jones, One-dimensional transport and the quantisation of the ballistic resistance. J. Phys. C Solid State Phys. **21**(8), L209 (1988)
10. K. Zimmermann, A. Jordan, F. Gay, K. Watanabe, T. Taniguchi, Z. Han, V. Bouchiat, H. Sellier, B. Sacépé, Gate-tunable transmission of quantum Hall edge channels in graphene quantum point contact. arXiv preprint arXiv:1605.08673, May 2016
11. H.L. Stormer, Nobel lecture: the fractional quantum Hall effect. Rev. Mod. Phys. **71**(4), 875 (1999)
12. T. Ando, A.B. Fowler, F. Stern, Electronic properties of two-dimensional systems. Rev. Mod. Phys. **54**(2), 437 (1982)
13. M. Büttiker, Quantized transmission of a saddle-point constriction. Phys. Rev. B **41**(11), 7906 (1990)
14. E. Scheer, P. Joyez, D. Esteve, C. Urbina, M.H. Devoret, Conduction channel transmissions of atomic-size aluminum contacts. Phys. Rev. Lett. **78**(18), 3535 (1997)
15. S. Frank, P. Poncharal, Z.L. Wang, W.A. de Heer, Carbon nanotube quantum resistors. Science **280**(5370), 1744–1746 (1998)
16. Z. Iftikhar, A. Anthore, S. Jezouin, F.D. Parmentier, Y. Jin, A. Cavanna, A. Ouerghi, U. Gennser, F. Pierre, Primary thermometry triad at 6 mK in mesoscopic circuits. Nature Commun. **7**, 12908 (2016)
17. R.B. Laughlin, Nobel lecture: fractional quantization. Rev. Mod. Phys. **71**(4), 863 (1999)
18. K. Pakrouski, M.R. Peterson, T. Jolicoeur, V.W. Scarola, C. Nayak, M. Troyer, Phase diagram of the ν= 5/2 fractional quantum hall effect: effects of Landau-level mixing and nonzero width. Phys. Rev. X **5**(2), 021004 (2015)
19. S.M. Girvin, The quantum hall effect: novel excitations and broken symmetries, in *Aspects Topologiques de la Physique en Basse Dimension. Topological Aspects of Low Dimensional Systems* (Springer, 1999), pp. 53–175
20. D.B. Chklovskii, B.I. Shklovskii, L.I. Glazman, Electrostatics of edge channels. Phys. Rev. B **46**(7), 4026 (1992)
21. C. Oriol, A. Martin. *Mécanismes inélastiques dans des circuits mésoscopiques réalisés dans des gaz bidimensionnels d'électrons*. Ph.D. thesis, Paris 11, 2010
22. G.-L. Ingold, Y.V. Nazarov, Charge tunneling rates in ultrasmall junctions, in *Single Charge Tunneling* (Springer, 1992), pp. 21–107
23. C. Altimiras, U. Gennser, A. Cavanna, D. Mailly, F. Pierre, Experimental test of the dynamical Coulomb blockade theory for short coherent conductors. Phys. Rev. Lett. **99**(25), 256805 (2007)
24. A.A. Odintsov, G. Falci, G. Schön, Single-electron tunneling in systems of small junctions coupled to an electromagnetic environment. Phys. Rev. B **44**(23), 13089 (1991)
25. P. Joyez, D. Esteve, Single-electron tunneling at high temperature. Phys. Rev. B **56**(4), 1848 (1997)
26. F.D. Parmentier, A. Anthore, S. Jezouin, H. Le Sueur, U. Gennser, A. Cavanna, D. Mailly, F. Pierre, Strong back-action of a linear circuit on a single electronic quantum channel. Nat. Phys. **7**(12), 935–938 (2011)

27. R. Cron, E. Vecino, Devoret M., Esteve, D., P. Joyez, A. L. Yeyati, C. Urbina, Dynamical Coulomb blockade in quantum point contacts, in *Electronic Correlations: from Meso- to Nanophysics, ed. by T. Martin, G. Montambaux, J. Trân Thanh Vân* (2001), p. 17
28. Ronald Cron. *Les contacts atomiques: un banc d'essai pour la physique mésoscopique*. Ph.D. thesis, Université Pierre et Marie Curie-Paris VI, 2001
29. Y.M. Blanter, M. Büttiker, Shot noise in mesoscopic conductors. Phys. Rep. **336**(1), 1–166 (2000)
30. K.G. Wilson, The renormalization group: critical phenomena and the Kondo problem. Rev. Mod. Phys. **47**(4), 773–840 (1975)
31. R. Bulla, T.A. Costi, T. Pruschke, Numerical renormalization group method for quantum impurity systems. Rev. Mod. Phys. **80**(2), 395 (2008)
32. K.M. Stadler, A.K. Mitchell, J. von Delft, A. Weichselbaum, Interleaved numerical renormalization group as an efficient multiband impurity solver. Phys. Rev. B **93**(23), 235101 (2016)
33. A.M. Tsvelick, P.B. Wiegmann, Solution of the n-channel Kondo problem (scaling and integrability). Z. für Phys. B Condens. Matter **54**(3), 201–206 (1984)
34. N. Andrei, C. Destri, Solution of the multichannel Kondo problem. Phys. Rev. Lett. **52**(5), 364 (1984)
35. A.M. Tsvelick, The transport properties of magnetic alloys with multi-channel Kondo impurities. J. Phys. Condens. Matter **2**(12), 2833 (1990)
36. I. Affleck, W.W. Ludwig, Critical theory of overscreened Kondo fixed points. Nucl. Phys. B **360**(2), 641–696 (1991)
37. I. Affleck, A.W. Ludwig, Exact conformal-field-theory results on the multichannel Kondo effect: single-fermion Green's function, self-energy, and resistivity. Phys. Rev. B **48**(10), 7297 (1993)
38. V.J. Emery, S. Kivelson, Mapping of the two-channel Kondo problem to a resonant-level model. Phys. Rev. B **46**(17), 10812 (1992)
39. K.A. Matveev, Coulomb blockade at almost perfect transmission. Phys. Rev. B **51**(3), 1743–1751 (1995)
40. A. Furusaki, K.A. Matveev, Theory of strong inelastic cotunneling. Phys. Rev. B **52**(23), 16676–16695 (1995)
41. L.I. Glazman, K.A. Matveev, Lifting of the Coulomb blockade of one-electron tunneling by quantum fluctuations. Sov. Phys. JETP **71**, 1031–1037 (1990)
42. N.C. van der Vaart, A.T. Johnson, L.P. Kouwenhoven, D.J. Maas, W. de Jong, M.P. de Ruyter, A. van Steveninck, C.J.P.M. van der Enden, Harmans, C.T. Foxon, Charging effects in quantum dots at high magnetic fields. Phys. B Condens. Matter **189**(1–4), 99–110 (1993)
43. C. Pasquier, U. Meirav, F.I.B. Williams, D.C. Glattli, Y. Jin, B. Etienne, Quantum limitation on Coulomb blockade observed in a 2D electron system. Phys. Rev. Lett. **70**(1), 69 (1993)
44. I.L. Aleiner, L.I. Glazman. Mesoscopic charge quantization. Phys. Rev. B **57**(16), 9608 (1998)
45. K.A. Matveev, A.V. Andreev, Thermopower of a single-electron transistor in the regime of strong inelastic cotunneling. Phys. Rev. B **66**(4), 045301 (2002)
46. S. Jezouin, Z. Iftikhar, A. Anthore, F.D. Parmentier, U. Gennser, A. Cavanna, A. Ouerghi, I.P. Levkivskyi, E. Idrisov, E.V. Sukhorukov, L.I. Glazman, F. Pierre, Controlling charge quantization with quantum fluctuations. Nature **536**(7614), 58–62 (2016)
47. J. Kondo, Resistance minimum in dilute magnetic alloys. Progress Theoret. Phys. **32**(1), 37–49 (1964)
48. P. Nozières, A "Fermi-liquid" description of the Kondo problem at low temperatures. J. Low Temp. Phys. **17**(1–2), 31–42 (1974)
49. P. Nozières, A. Blandin, Kondo effect in real metals. J. Phys. **41**(3), 19 (1980)
50. P.W. Anderson, A poor man's derivation of scaling laws for the Kondo problem. J. Phys. C Solid State Phys. **3**(12), 2436 (1970)
51. K.A. Matveev, Quantum fluctuations of the charge of a metal particle under the Coulomb blockade conditions. Sov. Phys. JETP **72**(5), 892–899 (1991)

52. E. Lebanon, A. Schiller, F.B. Anders. Coulomb blockade in quantum boxes. Phys. Rev. B **68**(4) (2003)
53. H. Yi, C.L. Kane, Quantum Brownian motion in a periodic potential and the multichannel Kondo problem. Phys. Rev. B **57**(10), R5579–R5582 (1998)
54. P. Nozières, Kondo effect for spin 1/2 impurity a minimal effort scaling approach. J. Phys. **39**(10), 8 (1978)
55. A.K. Mitchell, L.A. Landau, L. Fritz, E. Sela, Universality and scaling in a charge two-channel Kondo device. Phys. Rev. Lett. **116**(15) (2016)
56. Z. Iftikhar, A. Anthore, A.K. Mitchell, F.D. Parmentier, U. Gennser, A. Ouerghi, A. Cavanna, C. Mora, P. Simon, F. Pierre, Tunable quantum criticality and super-ballistic transport in a 'charge' kondo circuit, August 2017
57. B. Keimer, S.A. Kivelson, M.R. Norman, S. Uchida, J. Zaanen, From quantum matter to high-temperature superconductivity in copper oxides. Nature **518**(7538), 179–186 (2015)
58. S. Sachdev, B. Keimer, Quantum criticality (2011). arXiv preprint arXiv:1102.4628
59. M. Vojta, Quantum phase transitions. Rep. Prog. Phys. **66**(12), 2069 (2003)
60. H.T. Mebrahtu, I.V. Borzenets, D.E. Liu, H. Zheng, Y.V. Bomze, A.I. Smirnov, H.U. Baranger, G. Finkelstein, Quantum phase transition in a resonant level coupled to interacting leads. Nature **488**(7409), 61–64 (2012)
61. H.T. Mebrahtu, I.V. Borzenets, H. Zheng, Y.V. Bomze, A.I. Smirnov, S. Florens, H.U. Baranger, G. Finkelstein, Observation of Majorana quantum critical behaviour in a resonant level coupled to a dissipative environment. Nat. Phys. **9**(11), 732–737 (2013)
62. A.J. Keller, L. Peeters, C.P. Moca, I. Weymann, D. Mahalu, V. Umansky, G. Zaránd, D. Goldhaber-Gordon, Universal Fermi liquid crossover and quantum criticality in a mesoscopic system. Nature **526**, 237–240 (2015)
63. D.L. Cox, A. Zawadowski, Exotic Kondo effects in metals: magnetic ions in a crystalline electric field and tunnelling centres. Adv. Phys. **47**(5), 599–942 (1998)
64. M. Pustilnik, L. Borda, L.I. Glazman, J. von Delft, Quantum phase transition in a two-channel-Kondo quantum dot device. Phys. Rev. B **69**(11), 115316 (2004)
65. L. Saminadayar, D.C. Glattli, Y. Jin, B. Etienne, Observation of the e/3 fractionally charged Laughlin quasiparticle. Phys. Rev. Lett. **79**(13), 2526 (1997)
66. L.A. Landau, E. Cornfeld, E. Sela, Charge fractionalization in a kondo device, October 2017
67. T.K.T. Nguyen, M.N. Kiselev. Seebeck effect on a weak link between fermi and non-fermi liquids (2017)
68. C. Jayaprakash, H.R. Krishna-Murthy, J.W. Wilkins, Two-impurity Kondo problem. Phys. Rev. Lett. **47**(10), 737 (1981)
69. P. Gegenwart, Q. Si, F. Steglich, Quantum criticality in heavy-fermion metals. Nat. Phys. **4**(3), 186–197 (2008)

Chapter 2
Charge Quantization

In this chapter, we address the very basic problem of how the quantization of the charge as a multiple of the elementary electron charge[1] e evolves when an isolated conductor called *island* is progressively connected to an electron reservoir.

In our experiment, the island consists of a micron-size piece of metal. When the island is weakly connected to an electronic reservoir (e.g. by a tunnel junction), the electrons wave functions remain localized on the island and the charge quantization is preserved. As the connection to the reservoir is increased, due to quantum fluctuations, the wave functions spread out of the island and the charge quantization on the island is progressively reduced. Eventually, when the island is perfectly connected to a reservoir the number of charge localized on the island can no longer be defined: charge quantization is destroyed. Thus, having tunable and well-characterized contacts to the island provides a knob to control the degree of charge quantization. In practice this is achieved by using single electronic channels made in a semiconductor.

In the first section of this chapter we will present previous experimental and theoretical investigations on charge quantization. In particular, we will discuss conflicting experiments regarding the criterion to destroy completely charge quantization. The second section is devoted to the quantitative theoretical predictions on the degree of charge quantization versus the connection strength and the temperature. In the last section, we discuss experimental results on the control of charge quantization using quantum fluctuations.

[1]The value given by the NIST (http://physics.nist.gov/cgi-bin/cuu/Value?e) for this constant is: $e = -(1.6021766208 \pm 0.0000000098) \times 10^{-19} C$.

Z. Iftikhar, *Charge Quantization and Kondo Quantum Criticality in Few-Channel Mesoscopic Circuits*, Springer Theses,
https://doi.org/10.1007/978-3-319-94685-6_2

2.1 Previous Investigations on Charge Quantization

The quantization of the charge on an isolated system is quite an old topic. After the seminal experiment of Millikan in 1909, nanofabrication has allowed to build the first single-electron practical devices in the eighties. Relying on charge quantization, these single-electronics devices have applications in metrology, for charge detection or temperature measurement.

This section mixes the presentation of early experimental investigations and theoretical explanation of these observations. It is divided in two subsection: first, we discuss almost isolated systems; and second, we consider islands that are almost perfectly connected to an electrical circuit.

2.1.1 Quantization of the Charge in (Almost) Isolated Systems

We quickly present Millikan's experiment of isolated oil drops. Then we turn to nano-devices embedding almost isolated parts called *island* and explain the criteria to have a well quantized charge on the island.

Millikan's oil drop experiment

The most famous experiment on charge quantization is probably the seminal work of Millikan [1] led in 1909 where he measured speed of oil drops subject to a constant electric field $\vec{E}$ created between two electrodes. The electrostatic force $q\vec{E}$ directly depends on the excess charge q of the oil drop (which can be ionized with X-rays). It has been observed that the excess charge q is a multiple of a fundamental constant e which can be estimated from the knowledge of the viscosity of air and other parameters. In this experiment, the drops are isolated systems that can carry only an integer number of excess charge (see Fig. 2.1).

Conditions for a well-quantized charge state in nanostructures

Single-electron effects can be observed in nanofabricated devices that contain weakly connected parts called *islands*. The geometrical shape of an island determines its capacitance C. The smaller the island, the harder it is to add (or remove) an extra electron. Such an operation typically costs the charging energy $E_C \triangleq e^2/(2C)$. Charging effects are best visible when the thermal energy $k_B T$ and the voltage bias V_{dc} are negligible compared to the charging energy E_C:

$$k_B T, eV_{\mathrm{dc}} \ll E_C \tag{2.1}$$

In principle, charging effects remain measurable up to the charging energy E_C. A simple criterion on the weakness of the connection to ensure that charge is quantized is found in the following lines. The connection strength can be evaluated by the

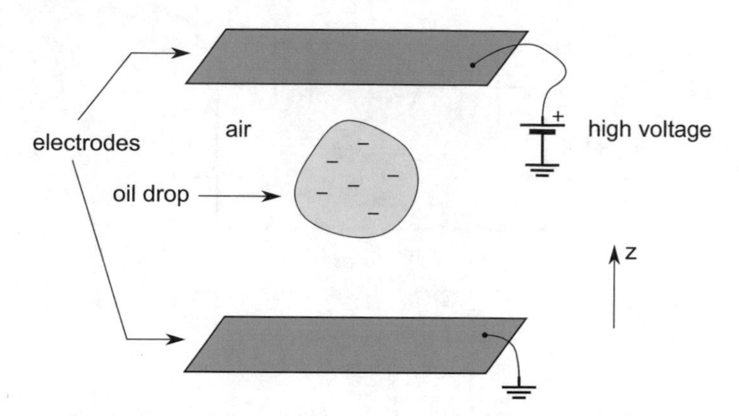

Fig. 2.1 A charged oil drop falling between two electrodes. An oil drop (in light gray) with five charges in excess (minus signs) is falling (see the vertical z-axis). The observation of its motion between two electrodes (in dark gray) can give access to the elementary charge e. This oil drop is an isolated system: its charge can only vary by amounts e

conductance G of the junction that connects the island to an electron reservoir. The typical time to discharge the capacitance C through this junction is $t_{\mathrm{RC}} = (1/G)C$. If the associated energy uncertainty $\Delta E \sim h/t_{\mathrm{RC}}$ becomes comparable to E_C (the typical energy cost to increment the charge on the island by one), then the charge state energy is ill-defined. For a well defined island charge, one needs $\Delta E \sim h/t_{\mathrm{RC}} \ll E_C = e^2/(2C)$, where the capacitance C can be simplified, and it comes:

$$G \ll G_K \triangleq e^2/h \tag{2.2}$$

Charge quantization is therefore destroyed by two types of fluctuations: (i) thermal fluctuation and (ii) quantum fluctuations.

Single electron transistor (SET)

The SET is a simple single-electron nano-device that consists of an island weakly connected to a circuit through two tunnel junctions (see Fig. 2.2a). As in a usual transistor, the conductance G_{SET} through the device can be modulated thanks to a voltage gate V_g. It has been extensively studied since its first realization in 1987 by Fulton and Dolan [2]. All the experiments we discuss in this section are based on a SET-like geometry.

Appendix B explains the physics of the SET. In particular, a quantitative description will be given, based on the perturbation theory (also known as the 'orthodox theory') that holds for $G \ll G_K$. This theory predicts 'Coulomb blockade oscillations' (see Fig. 2.2b) of the conductance versus a plunger gate voltage V_g that are characteristic of charge quantization (the number of electron localized on the island is incremented by one after each conductance peak).

The charge state of the island in SET can therefore be controlled at the single electron level. This offers possibilities for practical applications. For instance SET

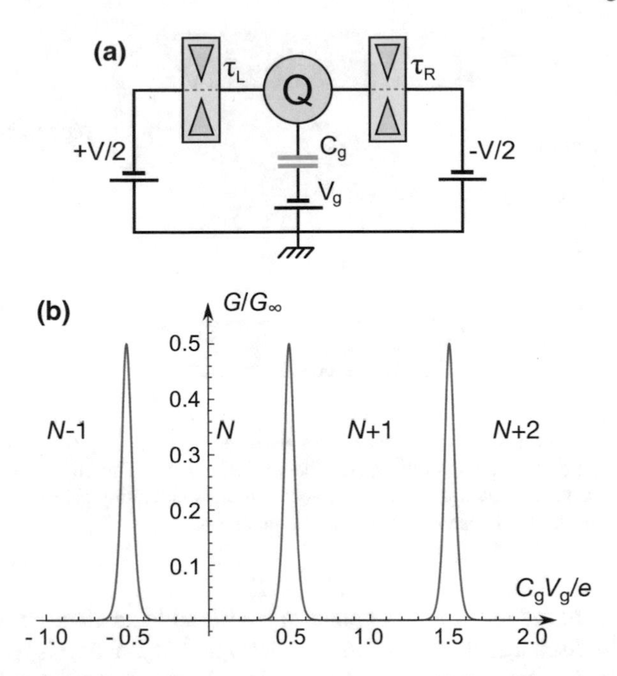

Fig. 2.2 Sigle electron transistor and Coulomb blockade oscillations. **a** Schematics of a SET. The device is d.c. biased with a voltage V. The excess charge Q on the island is quantized at low energy for weak transmissions $\tau_{L,R} \ll 1$ of the electronic channel (dotted lines). The charge state of the island can be tuned thanks to the voltage V_g applied on a gate coupled capacitively (C_g). **b** Theory predicts e-periodic oscillations of the conductance versus $C_g V_g$. This plot is made in the linear regime of small voltage bias, $T = 10\,\text{mK}$, and $E_C = 300\,\text{mK}$ (for details, see Fig. B.3). The number N of excess electrons on the island changes by one at each peak. Instead of V_g, we will rather use $N_g \triangleq C_g V_g / e$, the *continuous* control variable that tune the average number N of electron on the island (N is an integer when $\tau_{L,R} \ll 1$)

can be used as sensitive electrometers (see [3, 4] where a SET is used to probe the charge on a nearby island). Coulomb blockade in single-electron devices can also be used to perform primary thermometry (see [5], or our publication [6], both references are open access).

One should increase the conductance G of the connection between the island and the circuit to observe a destruction of charge quantization of the island of a SET. The intrinsic conductance of a junction can be increased by roughly two means: either one increases the conductance of a *single* electronic channel or one uses a wider tunnel junction (*many* tunnel channels in parallel, but none close to perfect transmission). Joyez and co-workers [7] have explored the second method with a metallic[2] island. They observed that the charge on the island can be quantized even when the conductance of many tunnel channels in parallel exceeds the quantum of conductance G_K. In this thesis, we focus on the first method, as we will see that even

[2]The metallic character of the island is of great importance as it will be explained below.

a *single* channel at perfect transmission completely destroys the charge quantization of a metallic island.

2.1.2 Coulomb Blockade at Almost Perfect Transmission

The question of Coulomb blockade close to perfect transmission has been addressed in some of the first implementations of SETs in the early nineties. However the criterion to completely destroy the quantization has been established in 1995 [8]. This subsection is divided into three parts: (i) we will discuss two experiments with contradictory conclusions on the complete destruction of charge quantization in the limit of a ballistic connection between the island and the circuit; (ii) we will present the theory that explain the two conclusions depending on whether coherent effects are considered or not (iii) we will show a recent experiment in presence of coherent effects where mesoscopic conductance oscillations are clearly observed beyond the ballistic limit.

Controversy on the absence of charge quantization in the limit of a ballistic connection between the island and the circuit

First of all, let me clarify that an electronic channel reaches the 'ballistic' limit as soon as there is no backscattering of electrons. In general, the conductance of such a channel is $G = G_K$, the quantum of conductance. At zero magnetic field, the conductance is quantized in units of $2 \times G_K$: because of spin degeneracy, pairs of two identical channels participate to transport. In presence of fractional quantum Hall effect, the ballistic limit is reached for a conductance $G = \nu G_K$, where the filling factor e.g. $\nu = 1/3$ can be lower than one. In other words, the criterion is not about the conductance itself, but rather about the transmission τ of the electronic channel defined as $\tau \triangleq G/G_{\text{ballistic}}$ where $G_{\text{ballistic}}$ is the conductance in the ballistic limit.

Kouwenhoven and co-workers studied charge quantization for different magnetic field when approaching the ballistic connection limit [9, 10]. Their sample can be represented by the SET shown in Fig. 2.2a, but with a non-metallic island. They have observed that Coulomb oscillations disappear as soon as the transmission of a channel becomes ballistic in the three regimes (at zero magnetic field, under a strong magnetic field to lift the spin degeneracy and with very strong magnetic field to reach FQHE with $\nu = 1/3$). They have also observed that the periodicity of the oscillations is independent of the magnetic field over the full range (10% of variation over 12 T). This period is directly related to the geometrical capacitance C_g (which does not change despite the compressible regions that appear due to the QHE [11]).

In contrast, Pasquier and co-workers have measured the same kind of sample, but they have observed Coulomb oscillations at zero magnetic field even beyond the ballistic limit (up to $\tau_L + \tau_R \approx 3 \times (2 \times G_K)/(2G_K)$, where τ_L and τ_R are the transmission through the left and right QPC respectively) [12].

The reason why these two experiment have contradictory conclusions might be explained by the theory of mesoscopic Coulomb blockade [13].

Coulomb blockade theory at the ballistic limit in presence or in absence of mesoscopic coherent effects

Matveev has proposed a theory [8] in order to explain the destruction of charge quantization observed in [10] when the transmission τ of a single channel through the QPC becomes perfect $\tau \longrightarrow 1$. He modeled the channel as a 1D conductor and used the bosonization technique to demonstrate that the Hamiltonian does not depend on the voltage gate V_g in the $\tau = 1$ ballistic limit. Note that, at this point, the Coulomb blockade problem is nonperturbative in the charging energy E_C and an exact treatment is needed [13]. At zero temperature, quantum fluctuations of the charge are known to appear when progressively connecting the island to a circuit [14] (see the dashed line in Fig. 2.3). Near the ballistic limit, Matveev has computed the charge Q on the island averaged on the quantum fluctuations (neglecting the thermal fluctuation) [8]:

$$\langle Q \rangle = eN_g - \gamma e/\pi \times |r| \sin 2\pi N_g \tag{2.3}$$

in the presence of a single (spinless) channel of small reflection coefficient amplitude $r \triangleq \sqrt{1-\tau} \ll 1$, and where the rescaled gate voltage N_g was defined in Fig. 2.2b. This expression shows quantitatively how the degree of charge quantization progressively reduces as $r \longrightarrow 0$ (see Fig. 2.3 for an illustration).

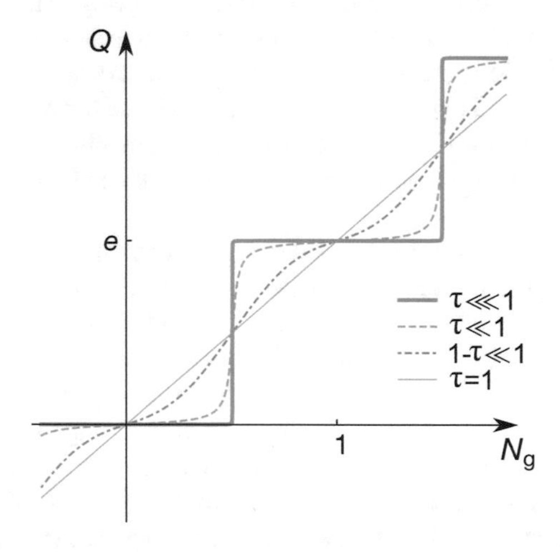

Fig. 2.3 Destruction of charge quantization by quantum fluctuations at zero temperature (reproduced from [8]). The average charge Q is plotted versus the gate voltage $N_g \triangleq C_g V_g/e$ at zero temperature for several reflection amplitudes $r \triangleq \sqrt{1-\tau}$. At $\tau \lll 1$, the island is almost isolated and Q is fully quantized (solid thick line). With a small finite transmission $\tau \ll 1$, quantum fluctuation smear the steps (dashed line). The weak backscattering regime $\sqrt{1-\tau} \ll 1$ is described by Eq. 2.3 (dash-dotted line). At perfect transmission $\tau = 1$, no modulation persists (thin line)

Table 2.1 Comparison of several experiment characteristics. Our experiment (last line) based on a *metallic* island with a negligible δE is given for comparison. Mesoscopic Coulomb blockade have been observed in all the listed experiments except in Kouwenhoven's one and in ours.

First author	T_{base} (mK)	$\delta E/k_B$ (mK)	E_C/k_B (K)	References
Kouwenhoven	10	120	7.0	[9]
Pasquier	60	85	2.2	[12]
Crouch	75	400	2.7	[15]
Cronenwett	100	163	3.3	[16]
Amasha	13	31	1.3	[4]
Jezouin	17	0.2	0.3	[17]

In his derivation, Matveev assumed that the mean energy level spacing δE in the island was negligible (compared to $k_B T$). In other words, he assumed that the density of state in the island was continuous. However, this is not exactly the case in the two experiments we mentioned above. A comparison of several sample characteristics is given in Table 2.1.

Aleiner and Glazman [13] have considered a finite level spacing δE and they have found that coherence effects lead to a persistence of charge quantization even in the ballistic limit $\tau = 1$! Indeed, the walls of the dot can reflect back the electrons into the channel (see Fig. 2.4). This elastic process depends on the path of the electron in dot. As for weak localization, interferences of electrons can modulate conductance through the dot [4]. This coherent effect leads to oscillations of the differential capacitance $C_{\text{diff}} \triangleq \mathrm{d}\langle Q \rangle / \mathrm{d}V_g$ that have the same period as the 'usual' Coulomb oscillations and an amplitude of $\sqrt{\delta E/E_C}$ for spinless electrons and $(\delta E/E_C)\ln^2(\delta E/E_C)$ for spin-half electrons.

These oscillations that may occur in *quantum dots* (islands with a discrete density of state) even when they are open (connected to a lead with a channel of ballistic transmission $\tau = 1$) are called 'mesoscopic Coulomb blockade' oscillations. As they require phase coherence, they highly depend on temperature and magnetic field.

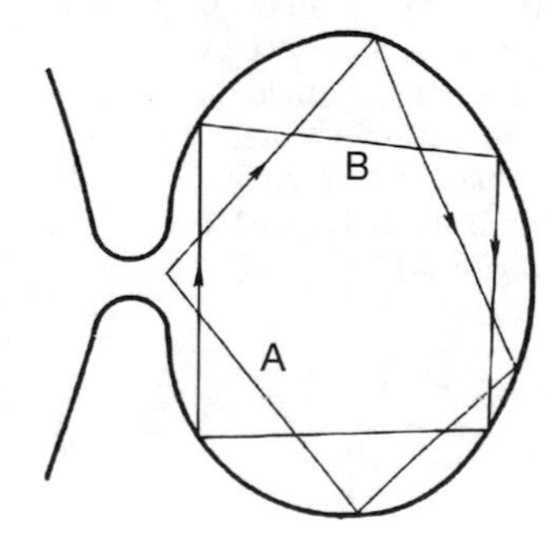

Fig. 2.4 Mesoscopic effect in a quantum dot (reproduced from [13]). In some experiments (not in ours), electrons can be coherently reflected back to the electronic reservoir (on the left) leading to a modulation of the differential capacitance that depends on the trajectory (e.g. A or B)

Mesoscopic Coulomb blockade observations

In principle, mesoscopic Coulomb blockade should have been observed in the experiment of Kouwenhoven and co-workers [9]. Indeed, in the SET regime, they have observed that the conductance through the two tunnel junctions ($\tau_1, \tau_2 \ll 1$) in series was much larger ($G_{\mathrm{SET}} \approx 0.8 G_K$) than the classical serial conductance. At low temperature T compared to the level spacing $k_B T \ll \delta E$, the system enters the quantum regime [18] where coherent effects are relevant. According to Aleiner and Glazman [13], a possible reason why Kouwenhoven and co-workers have not observed mesoscopic Coulomb blockade is that the shape of their quantum dot might not be suitable to host the chaotic paths required by the theory.

After the experiment of Pasquier and co-workers [12], other teams have observed a persistence of quantization in the limit of a ballistic channel connection [4, 15, 16]. They have measured a temperature dependence on these Coulomb oscillations. In [4, 16], a strong dependence on the magnetic field have been observed (see Fig. 2.5). These two dependences are characteristic of mesoscopic phenomena as they involve the quantum phase of the electrons.

In our implementation, we use a metallic island where δE is six orders of magnitude smaller than E_C and thus avoid any mesoscopic Coulomb blockade. That is

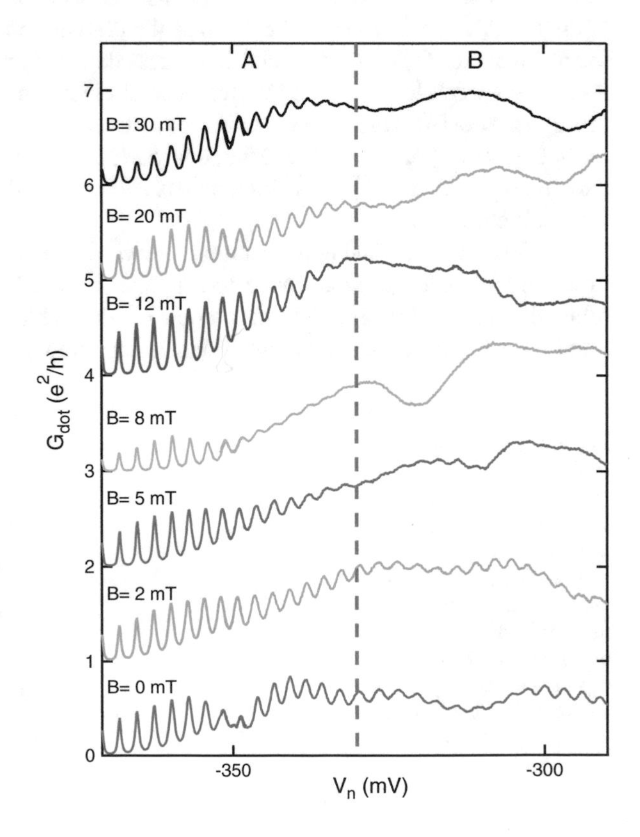

Fig. 2.5 Mesoscopic Coulomb blockade observation by Amasha and co-workers [4]. For several magnetic field B, the conductance G_{dot} is plotted versus a voltage gate V_n that controls the opening of the dot. In region A, the dot is partially connected to the circuit; whereas is region B, the dot is open (an electronic channel is fully transmitted). Conductance oscillations are observed in region B, they are strongly affected by a finite magnetic field (in contrast to the 'usual' Coulomb blockade oscillations observed in region A)

why our sample specifically probes (usual) Coulomb blockade without any coherent effect 'artifact'. Theoretical predictions on how charge quantization vanishes in this case is presented in the following section.

2.2 Theoretical Predictions on the Charge Quantization of a Metallic Island Without Coherent Effect

In addition to avoid mesoscopic Coulomb blockade, considering a metallic island without coherent effect has the advantage to simplify the problem for the theorists. In this section we will give some *quantitative* prediction on the conductance G_{SET} at zero voltage bias $V = 0$ (see Fig. 2.2a) in the quantum regime (low temperature $k_B T \ll E_C$) in two limit cases: the asymmetric case $\tau_L \ll 1$, $1 - \tau_R \ll 1$ and the symmetric case $1 - \tau_{L,R} \ll 1$.

We will first explain how both the conductance G_{SET} and the differential capacitance C_{diff} characterize the charge quantization. Then we will present the predictions in the two limit cases (asymmetric and symmetric). For each case, we will separate the low temperature quantum regime and the high temperature regime. The latter regime where many charge states are thermally activated has been treated by our collaborators Levkivskyi, Idrisov, Sukhorukov and Glazman to explain our experimental data (shown in the next section). We will see that all the predictions scales with reflection coefficient amplitude $\sqrt{1-\tau}$.

2.2.1 *Transport Versus Thermodynamic Properties*

The quantized aspect of the charge on the island can be probed either through transport or thermodynamic properties. The former consists in the measurement of the conductance G_{SET} of the elements QPC_L-island-QPC_R in series. The latter is based on the differential capacitance C_{diff} to the island.

We have not access to the differential capacitance C_{diff} with our sample because it requires an additional electrometer (for example, see [4] where two quantum dots have been used: a big one as an island, and a small one as an electrometer). This quantity $C_{\mathrm{diff}} \triangleq \mathrm{d}\langle Q\rangle/\mathrm{d}V_g$ is directly related to the average charge $\langle Q\rangle$ on the island. Note that a single lead can be used when studying C_{diff} since the current through the device is not considered when dealing with thermodynamic properties.

The e-periodic oscillations of the conductance G_{SET} versus the plunger gate V_g that we measured are also a clear signature of charge quantization. However, in contrast with C_{diff}, the measure of G_{SET} does not give a direct access to the average excess charge Q on the island. Nonetheless, we will show that both G_{SET} and C_{diff} have similar dependences on the relevant parameters. Moreover, the conductance is important in the applications of single-electronics [19].

2.2.2 Asymmetric Case $\tau_L \ll 1$, $1 - \tau_R \ll 1$

Low temperatures $k_B T \ll E_C$

The asymmetric limit has been studied by Matveev when he showed that charge quantization vanishes when an electronic channel connects ballistically the island [20]. We can find the visibility of the oscillations of the differential capacitance C_{diff} from his expression of the average charge given in Eq. 2.3:

$$\Delta C_{\text{diff}}(\tau_L \ll 1, 1 - \tau_R \ll 1) \triangleq \frac{C_{\text{diff}}^{\max} - C_{\text{diff}}^{\min}}{C_{\text{diff}}^{\max} + C_{\text{diff}}^{\min}} = 2\gamma\sqrt{1 - \tau_R} \tag{2.4}$$

where $\gamma \approx \exp(0.5772)$ is the exponential of Euler's constant. The quantitative prediction for the serial conductance through the two asymmetric junctions is given by (Eq. 34 in [21]):

$$G_{\text{SET}}(\tau_L \ll 1, 1 - \tau_R \ll 1, N_g) = \tau_L G_K \times \frac{2\pi^4 T^2}{3\gamma^2 E_C^2}\left[1 - 2\gamma\xi\sqrt{1 - \tau_R}\cos(2\pi N_g)\right] \tag{2.5}$$

where $\xi \approx 1.59$ is a numerical coefficient. This expression is valid only for $1 - \tau_R \ll 1$ (as one can see that the minimal value of the conductance in Eq. 2.5 is strikingly incorrect, negative, already for $\tau_R \leq 1 - 1/(2\gamma\xi)^2 \approx 0.97$). The visibility of the Coulomb blockade oscillations of the conductance comes directly from the quantitative expression of G_{SET}:

$$\Delta Q(\tau_L \ll 1, 1 - \tau_R \ll 1) \triangleq \frac{G_{\text{SET}}^{\max} - G_{\text{SET}}^{\min}}{G_{\text{SET}}^{\max} + G_{\text{SET}}^{\min}} = 2\gamma\xi\sqrt{1 - \tau_R} \tag{2.6}$$

We see that in the case of asymmetric junctions, at low temperature $T \ll E_C$, the visibility on the oscillations of the differential capacitance C_{diff} is proportional to the visibility on the conductance oscillations G_{SET}: $\Delta Q = \xi \Delta C_{\text{diff}}$, as explicitly pointed out by Yi and Kane in [22].

High temperature limit $k_B T \gg E_C/\pi^2$

At high temperatures $k_B T \gg E_C$, several charge states are populated. This situation has not been considered by Matveev in [8]. However, his reasoning can be extended to this case by averaging over the Gibbs distribution of fluctuations (see the Methods section of our article [17]). Our collaborators have then derived the expression of the differential capacitance at high temperature, in the asymmetric limit:

$$C_{\text{diff}}(\tau_L = 0, 1 - \tau_R \ll 1, N_g) \approx \frac{e}{\Delta V_g} - 4\frac{e}{\Delta V_g}\frac{\pi^2 k_B T}{E_C}\exp\left(-\frac{\pi^2 k_B T}{E_C}\right)\sqrt{1 - \tau_R}\cos(2\pi N_g)$$

This expression has been derived using also [23], which is a general theory not assuming $1 - \tau_R \ll 1$ at large temperatures $k_B T \gg E_C/\pi^2$. This expression matches the prediction for an almost isolated island ($\tau_{L,R} \ll 1$), just by taking the limit $\sqrt{1 - \tau_R} \longrightarrow 1$. The visibility on these C_{diff} oscillations is

$$\Delta C_{\text{diff}}(\tau_L = 0, 1 - \tau_R \ll 1) \approx 4\frac{\pi^2 k_B T}{E_C} \exp\left(-\frac{\pi^2 k_B T}{E_C}\right)\sqrt{1 - \tau_R}, \tag{2.7}$$

it is exponentially reduced at high temperatures and it depends as $\sqrt{1 - \tau_R}$ with the transmission. The conductance in the asymmetric and high temperature regime can be obtained starting from Eq. 2.5 in [24] or by using a specific method (see the Methods section of our article [17] and the unpublished ref. [25] of our collaborators). We get a visibility on G_{SET} oscillations as:

$$\Delta Q(\tau_L \ll 1, 1 - \tau_R \ll 1) \approx \exp\left(-\frac{\pi^2 k_B T}{E_C}\right)\sqrt{1 - \tau_R} \tag{2.8}$$

The same dependence on the transmission τ_R is predicted for the visibility of both the C_{diff} and G_{SET} oscillations. And an exponential reduction with the temperature is also expected at high temperature for the conductance oscillations.

2.2.3 Symmetric Limit $1 - \tau_{L,R} \ll 1$

Low temperatures $k_B T \ll E_C$

The conductance in the strong coupling limit $1 - \tau_{L,R} \ll 1$ with two channels almost perfectly transmitted has been studied by Furusaki and Matveev [24]. They have derived the quantitative expression for the serial conductance (see Eq. 38 obtained in [24] and Equation A9 for the T-linear term which is the leading-order correction in $k_B T/E_C$):

$$\begin{aligned} G_{\text{SET}}(1 - \tau_L \ll 1, 1 - \tau_R \ll 1, N_g) = G_K/2 \times \Bigg[1 - \\ \frac{\gamma E_C \Gamma_+(\tau_L, \tau_R, N_g)}{\pi^3 k_B T}\psi'\left(1/2 + \frac{\gamma E_C \Gamma_+(\tau_L, \tau_R, N_g)}{\pi^3 k_B T}\right) \\ - \frac{\pi^3 \gamma k_B T}{16 E_C} \times \Gamma_-(\tau_L, \tau_R, N_g)\Bigg] \end{aligned} \tag{2.9}$$

where we have replaced the original integral by the digamma function[3] and where

$$\Gamma_{\pm}(\tau_L, \tau_R, N_g) = (1 - \tau_L) + (1 - \tau_R) \pm 2\sqrt{(1 - \tau_L)(1 - \tau_R)} \cos(2\pi N_g) \quad (2.10)$$

The visibility of the Coulomb blockade oscillations at low temperature $k_B T / E_C \ll 1$ and in the strong coupling reads:

$$\Delta Q(1 - \tau_{L,R} \ll k_B T / E_C \ll 1) = \frac{\gamma E_C}{\pi k_B T} \sqrt{(1 - \tau_L)(1 - \tau_R)} \quad (2.11)$$

The differential capacitance has been evaluated at low temperature in the strong couping regime when first $\tau_L \longrightarrow 1$ and then $\tau_R \longrightarrow 1$ (Eq. 41 in [26]):

$$C_{\text{diff}}(1 - \tau_L \ll 1 - \tau_R \ll 1, N_g) = \frac{e}{\Delta V_g} - 4\gamma \frac{e}{\Delta V_g} \ln(1 - \tau_L) \sqrt{(1 - \tau_L)(1 - \tau_R)} \cos(2\pi N_g) \quad (2.12)$$

It yields a visibility on C_{diff} oscillations as:

$$\Delta C_{\text{diff}}(1 - \tau_L \ll 1 - \tau_R \ll 1) = 4\gamma \ln(1 - \tau_L) \sqrt{(1 - \tau_L)(1 - \tau_R)} \quad (2.13)$$

In this regime also, a dependence in $\sqrt{(1 - \tau_L)(1 - \tau_R)}$ is found for the visibility on both the G_{SET} and C_{diff} oscillations. However, a 'log' appears in the expressions of ΔC_{diff}.

Note that for exactly symmetric transmissions $\tau \triangleq \tau_L = \tau_R$, the differential capacitance diverges at the degeneracy point $N_g = 0$ (Eq. 49 in [8] or Eq. 32 in [26]):

$$C_{\text{diff}}(1 - \tau \ll 1, N_g) = \frac{e}{\Delta V_g} - 4\gamma \frac{e}{\Delta V_g} \ln\left((1 - \tau) \sin(\pi N_g)\right) \times (1 - \tau) \cos(2\pi N_g)$$

This divergence is due to the two-channel Kondo effect, which will be the topic of Chap. 3. It disappears as soon as an asymmetry $\tau_L \neq \tau_R$ is introduced [26]. This critical phenomenon will be discussed in Chap. 4.

High temperature regime $k_B T \gg E_C/\pi^2$

As for the asymmetric case, our collaborators have computed the conductance in the high temperature regime. They have found an expression that gives a visibility of:

$$\Delta Q(1 - \tau_{L,R} \ll 1) \approx \exp\left(-\frac{\pi^2 k_B T}{E_C}\right) \sqrt{1 - \tau_L} \sqrt{1 - \tau_R} \quad (2.14)$$

[3] We have verified numerically that $\frac{1}{4T} \int_{-\infty}^{\infty} \mathrm{d}E \frac{1}{\cosh^2(E/(2k_B T))} \frac{\Gamma_+^2(\tau_L, \tau_R, N_g)}{E^2 + \Gamma_+^2(\tau_L, \tau_R, N_g)} = \frac{\Gamma_+(\tau_L, \tau_R, N_g)}{2\pi k_B T} \psi' \left(1/2 + \frac{\Gamma_+(\tau_L, \tau_R, N_g)}{2\pi k_B T}\right)$.

In this strong coupling regime also, we see that Coulomb blockade oscillations on the conductance are exponentially reduced at high temperatures. The same behavior is expected for the visibility of the differential capacitance oscillation (this quantity has not been computed by our collaborators, but no interplay between the channels will occur out of the quantum regime $k_B T \ll E_C$).

2.2.4 Universality in the High Temperature Regime

The predictions beyond the tunnel regime require advanced theoretical methods. In particular, as we will see in Chap. 3, a reason is that the Coulomb blockade with a metallic island and a few electronic channels can exhibit Kondo effect at low temperature, depending on the gate voltage V_g.

The predictions we have presented above are based on the bosonization technique. Another method consists in using the instanton solution of Korshunov [27]. These instantons have been used in [28] to predict an exponential suppression of the charge quantization with the temperature. This technique has been used also by Nazarov to propose a general solution of the Coulomb blockade problem beyond the tunnel limit [23].

His solution is presented for thermodynamic quantities (such as C_{diff}) only. The charging energy E_C is shown to have effective modulations due to the gate voltage V_g that scale as $\tilde{E}_C \propto E_C \prod_j \sqrt{1-\tau_j}$, where the product is done on the electronic channel index j. This expressions suggests that the suppression of the charge quantization becomes universal when rescaled with respect to $\sqrt{1-\tau}$.

This prediction is not valid at low temperature[4] as it does not predict the log divergence on C_{diff} for symmetric transmissions near the ballistic limit (see Eq. 2.12 and Sect. 2.2.3).

2.3 Controlling Charge Quantization with Quantum Fluctuations

We have published our experimental data on the charge quantization of a metallic island on August 4th 2016 in Nature journal (https://doi.org/10.1038/nature19072). This section reproduces the figures of the paper and summarizes its content; for details, the interested reader can refer to the publication [17].

[4]Probably because of the large fluctuations of the phase at low temperature and $\sqrt{1-\tau} \ll 1$ that makes the instanton technique less accurate (private discussion with our collaborators).

2.3.1 Tunable Quantum Connection to a Metallic Island

A micrograph of the sample[5] is provided in Fig. 2.6a. This corresponds to the equivalent schematics shown in Fig. 2.6b. This sample is described and characterized in details in Appendix A. Its key features are (i) a metallic island with a negligible energy level spacing (see 118 for numbers) and (ii) highly tunable and characterizable QPCs (see Fig. 2.6c) that we use to adjust the connection between the island and the circuit. As the island is connected via a single electronic channels, the time an electron spends on the island (which is typically the invert of the energy level spacing in the island) is too large compared to its quantum lifetime. This avoids any coherent electronic transport from a reservoir on one side of the island to another.

The connection to the island is characterized on each side (left and right) by the transmission probability $\tau_{L,R}$ of a single electronic channel. For the completeness of the experiment, we would open a second electronic channel (e.g. on the right side). Due to the strong magnetic field, the second channel opens only once the first one is fully transmitted. Therefore, one can generalize the probability to values higher than one, for example: $\tau_R = 1.36$ means the first (outermost quantum Hall edge) channel is perfectly transmitted, and the second one has a probability of 0.36 to be transmitted.

For small transmissions $\tau_{L,R} \ll 1$, the circuit Fig. 2.6b is equivalent to a SET. In this regime, one could d.c. bias the sample to characterize the charging energy E_C of the island. In Fig. 2.6d, two Coulombs diamonds are displayed. Their height gives a charging energy $E_C \triangleq e^2/2C \approx 25\,\mu\text{eV} \approx k_B \times 0.3\text{K}$.

2.3.2 Charge Quantization Versus Connection Strength at $T \approx 17\,\text{mK}$

We study the charge quantization on the metallic island by measuring the conductance G_{SET} through "QPC$_L$-island-QPC$_R$" as a function of the voltage V_g applied on a plunger gate (in golden color in Fig. 2.6a and b). At low temperature $T \approx 17\,\text{mK} \ll E_C/k_B$, the charge quantization on the metallic island is visible on the e/C_g-periodic oscillations of $G_{\text{SET}}(V_g)$, where C_g is the geometrical capacitance of the plunger gate where V_g is applied.

In Fig. 2.7a, we plot $G_{\text{SET}}(V_g)$ for a fixed value of $\tau_L = 0.24$ and several values of $\tau_R \in \{0.1, 0.6, 0.88, 0.98, 1.5\}$. One can see that oscillations are visible for the four first graphs, and they disappear in the last one, where $\tau_R = 1.5 > 1$. For a more systematic study, one may consider the visibility of G_{SET} oscillations defined by $\Delta Q \triangleq (G_{\text{SET}}^{\max} - G_{\text{SET}}^{\min})/(G_{\text{SET}}^{\max} + G_{\text{SET}}^{\min})$. This is done in Fig. 2.7b, where we plot the visibility ΔQ as a function of τ_R, for several values of τ_L (see the legend), and where

[5] The shown sample is actually not the one used for the experiment but both belong to the same batch and are essentially the same. Indeed, the relatively important electron beam required to acquire such a high-resolution image might change the properties of the sample and introduce possible artifacts.

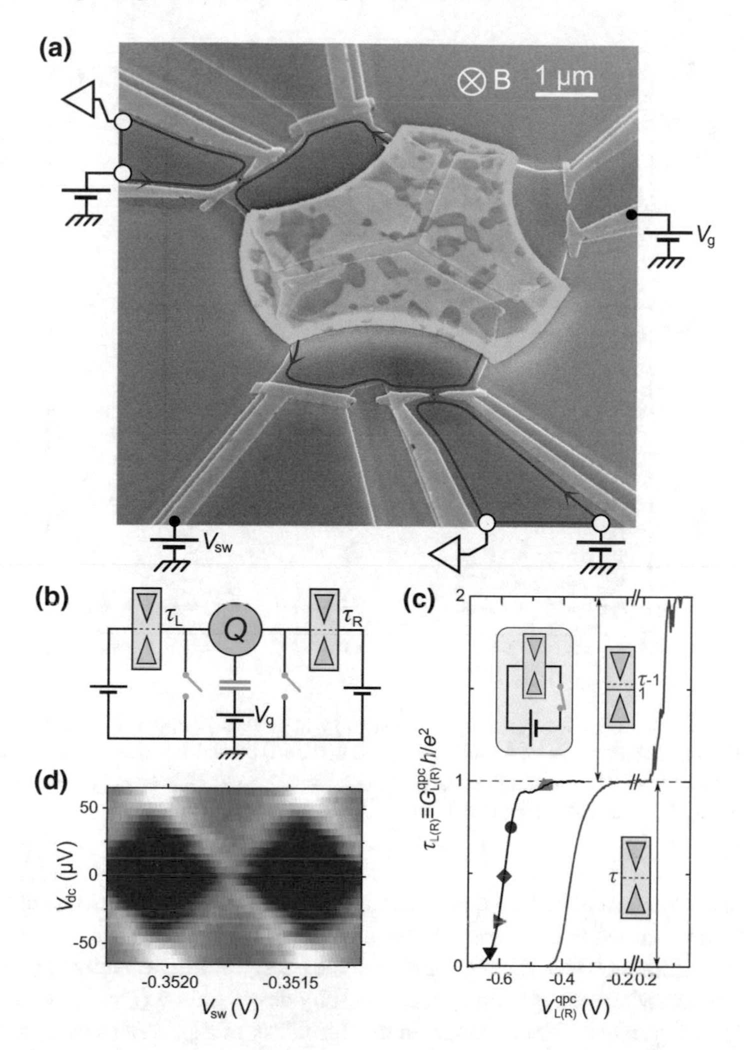

Fig. 2.6 Tunable quantum connection to a metallic island. **a** Colored sample micrograph. A micrometer-scale metallic island (red) is connected to large electrodes (white circles) through two quantum point contacts (QPCs, green split gates) formed in a buried 2D electron gas (2DEG, darker gray delimited by bright lines). The lateral gates (blue) implement short-circuit switches as shown in **b**. The top-right yellow gates, tuned at V_g negative enough to deplete the 2DEG underneath, are capacitively coupled to the island. In the applied $B \simeq 4$ T, the current propagates along two edge channels (red lines, that the represent only the outermost channel) in the direction indicated by arrows. **b** Sample schematic. **c** The 'intrinsic' (i.e. switch closed) conductance $G^{\mathrm{qpc}}_{\mathrm{L(R)}}$ across the top-left QPC$_L$ (bottom-right QPC$_R$) is shown versus split gate voltage $V^{\mathrm{qpc}}_{\mathrm{L(R)}}$ as a black (red) line. Symbols indicate the set-points of QPC$_L$ used thereafter. **d** Coulomb diamond patterns in the device conductance G_{SET} (larger shown brighter, from 0 in dark blue up to $0.13e^2/h$ in white) measured versus gate (V_{sw}) and bias (V_{dc}) voltages for tunnel contacts ($\tau_{L,R} \ll 1$)

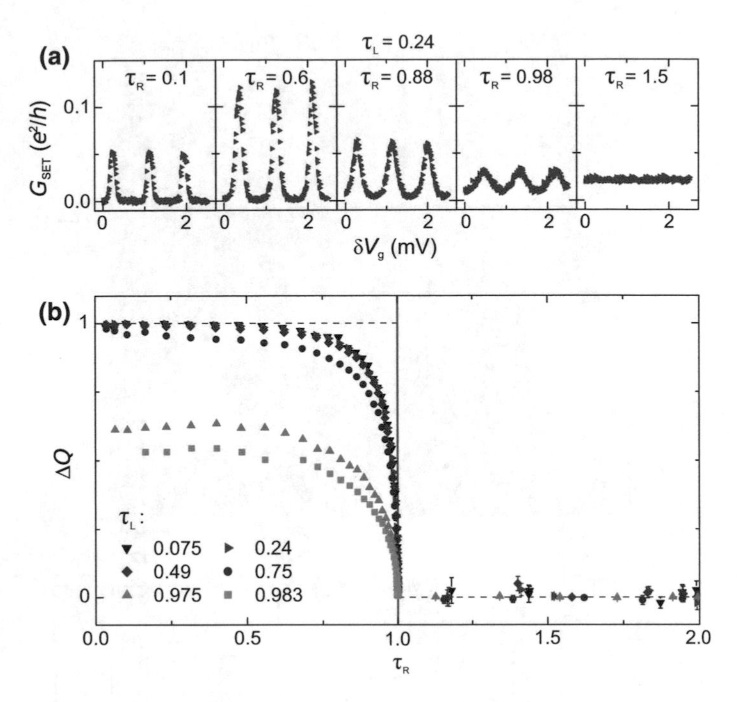

Fig. 2.7 Charge quantization versus connection strength at $T \approx 17\,\text{mK}$. **a** Conductance sweeps $G_{\text{SET}}(\delta V_g)$ with a fixed $\tau_L = 0.24$, and $\tau_R = 0.1, 0.6, 0.88, 0.98$ and 1.5, from left to right respectively. **b** Visibility of G_{SET} oscillations $\Delta Q \triangleq (G_{\text{SET}}^{\text{max}} - G_{\text{SET}}^{\text{min}})/(G_{\text{SET}}^{\text{max}} + G_{\text{SET}}^{\text{min}})$ versus τ_R, with each set of symbols corresponding to a different QPC$_L$ set-point

we see a destruction of charge quantization as soon as a single electronic channel is perfectly transmitted (i.e. without backscattering).

The transition between the regime where the charge is fully quantized ($\Delta Q \approx 1$) and the region where charge quantization is fully destroyed ($\Delta Q = 0$) is very sharp at this low temperature. For instance, in the third inset of Fig. 2.7a (where $\tau_L = 0.24$ and $\tau_R = 0.98$), whereas one is very close to the critical point $|1 - \tau_R| = 2 \times 10^{-2}$, we still observe large oscillations that correspond to a visibility $\Delta Q \approx 0.5$!

2.3.3 *Charge Quantization Scaling Near the Ballistic Critical Point*

In the previous subsection, we demonstrated the criterion of the destruction of charge quantization above the limit of one channel ballistically transmitted to a *metallic* island, as it was expected by Matveev's theory [8]. Let us now observe how the charge quantization is destroyed in the vicinity of the critical point $\tau = 1$.

At low temperature, the power law $\Delta Q \propto \sqrt{1 - \tau_R}$ is expected for $1 - \tau \ll 1$, when either $\tau_L \ll 1$ (see Eq. 2.5) or $1 - \tau_L \ll 1$ (see Fig. 2.9). In Fig. 2.8, we plot the same data as in Fig. 2.7, but in a log-log scale, and versus $1 - \tau_R$ (only the points with $1 - \tau_R > 2 \times 10^{-3}$ are displayed). This representation shows that in the vicinity of the critical point (when $1 - \tau_R < 0.02$), the charge quantization vanishes as $\sqrt{1 - \tau_R}$

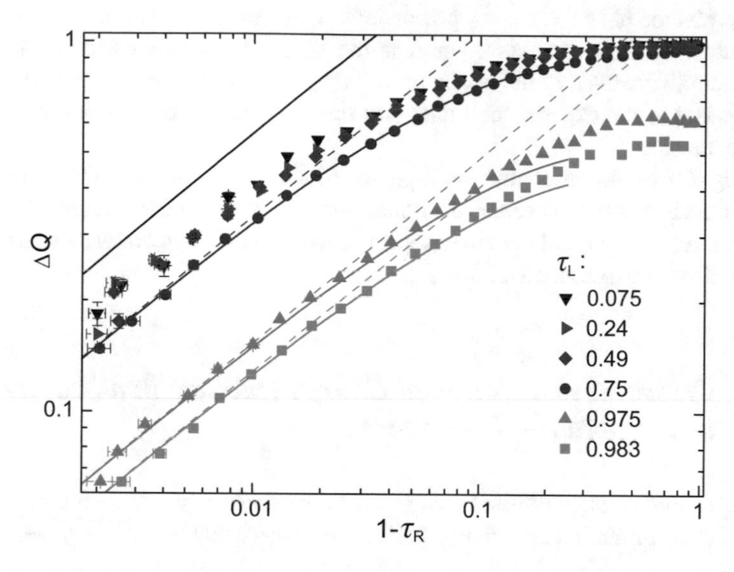

Fig. 2.8 Charge quantization scaling near the ballistic critical point. The ΔQ data at $T \simeq 17$ mK are displayed versus $1 - \tau_R$ in a log-log scale, with distinct sets of symbols for the different QPC$_L$ set-points. Continuous lines are quantitative predictions (no fit parameters) derived assuming $k_B T \ll E_C$, $1 - \tau_R \ll 1$, and either $\tau_L \ll 1$ (top continuous line) or $1 - \tau_L \ll 1$ (three bottom continuous lines). The power law $\Delta Q \propto \sqrt{1 - \tau_R}$ (straight lines) is systematically observed at $1 - \tau_R \lesssim 0.02$, also at intermediate τ_L

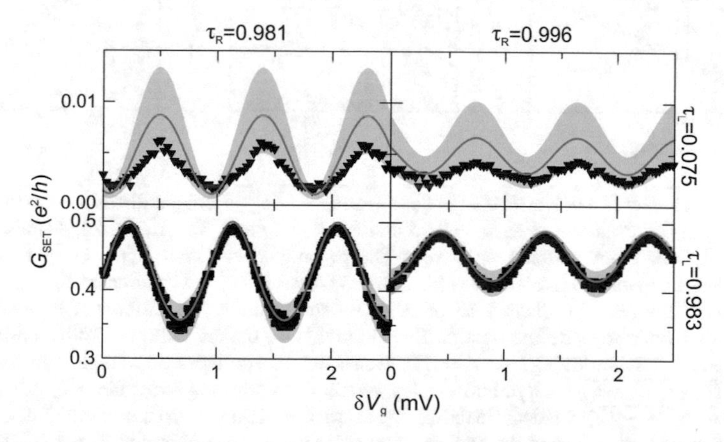

Fig. 2.9 Conductance measurements versus quantitative predictions Direct $G_{\mathrm{SET}}(\delta V_g)$ comparison at $T = 17$ mK between data (symbols) and predictions (solid lines, gray areas correspond to the temperature uncertainty of ± 4 mK) in the two limits addressed by theory (Eq. 2.5 for $\tau_L \approx 0$ (top panels), Eq. 2.9 for $\tau_L \approx 1$ (bottom panels)).

whatever the value of τ_L in the full range $\tau_L \in [0, 1]$, and not only in the two limits established theoretically.

Moreover, one can compare the experimental data to the quantitative theory of Eqs. 2.5 and 2.9. In the asymmetric limit $\tau_L \ll 1$, a mismatch of 25% is observed in the prefactor of the $\sqrt{1-\tau_R}$ power law. This could be explained by the finite temperature of the experiment, whereas the theory is exactly established at zero-temperature. However, in the opposite $1-\tau_{L,R} \ll 1$ limit, we notice a quantitative match between the experimental data and the low-temperature theory without any fitting parameter.

In Fig. 2.9 we compare the raw G_{SET} conductance oscillations (from which we extract the visibility ΔQ) versus the quantitative theory in the two known limits. We observe a relatively good agreement with respect to the temperature uncertainty in the quantitative expressions of G_{SET}.

2.3.4 *Crossover to a Universal Charge Quantization Scaling as Temperature Is Increased*

As temperature rises, additional charge states are thermally activated and this averages out charge quantization. In Fig. 2.10a, we observe that the $\sqrt{1-\tau_R}$ dependence

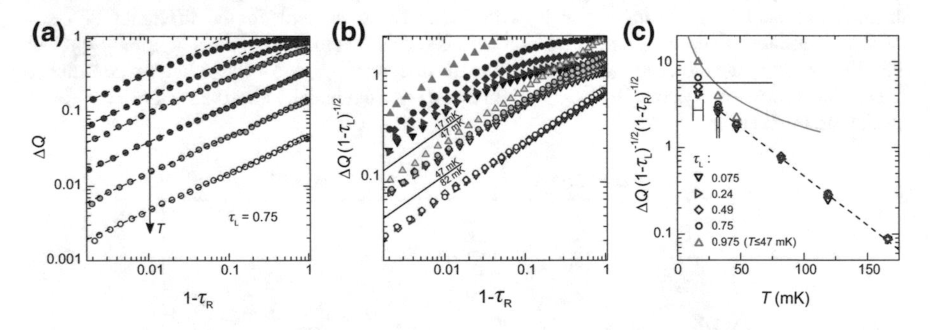

Fig. 2.10 Crossover to a universal charge quantization scaling as temperature is increased. **a** Symbols display ΔQ versus $1-\tau_R$ at $\tau_L = 0.75$ and for $T \simeq 17$ mK, 32 mK, 47 mK, 82 mK, 119 mK and 166 mK, from top to bottom respectively. The τ_R range over which $\Delta Q \propto \sqrt{1-\tau_R}$ (straight lines) extends up to the full interval $\tau_R \in [0, 1]$ when increasing T. **b** The rescaled $\Delta Q/\sqrt{1-\tau_L}$ is shown versus $1-\tau_R$, with distinct set of symbols corresponding to different QPC_L set-points as in **c**. Solid lines separate the data at $T \simeq 17$ mK (top, darker filling), 47 mK (middle) and 82 mK (bottom, brighter filling). At $T = 82$ mK, all the data collapse on a single universal curve $\Delta Q \propto \sqrt{(1-\tau_L)(1-\tau_R)}$. **c** Symbols display versus T, in semi-log scale, the fully rescaled data $\Delta Q/\sqrt{(1-\tau_L)(1-\tau_R)}$, extracted in the regime of small enough $1-\tau_R$ such that $\Delta Q \propto \sqrt{1-\tau_R}$. Horizontal error bars represent the experimental temperature uncertainty at $T = 17 \pm 4$ mK and 32 ± 1 mK. Solid lines are the quantitative predictions in the quantum regime $k_B T \ll E_C$, given by Eqs. 2.5 (black) and 2.9 (green). The straight dashed line displays an exponential decay close to predictions in the presence of strong thermal fluctuations (see text)

of the charge quantization ΔQ is more robust at higher temperature and this power law is eventually valid on the full range $\tau_R \in [0, 1]$ at high enough temperature.

Another observation is that at high temperature, the prefactor of the power law $\Delta Q \propto \sqrt{1-\tau_R}$ progressively becomes independent of τ_L. Above 82 mK , the destruction of charge quantization becomes fully universal (independent of τ_L and τ_R, see Fig. 2.10b and c). This regime, dominated by thermal fluctuations rather than quantum fluctuations of charge, is described by an exponential decay of $\Delta Q/\sqrt{1-\tau_L}/\sqrt{1-\tau_R} \propto \exp(-\alpha\pi^2 k_B T/E_C)$, where $\alpha \approx 0.8$ is a numerical factor to fit the experimental data shown in Fig. 2.10c. In this figure, one can observe a crossover from the quantum regime (where it exists a subtle interplay between τ_L and τ_R) to the thermal regime which is universal, as expected by the theory.

2.4 Conclusion

The use of a hybrid metal-semiconductor device allowed us to probe and control the degree of charge quantization in the metallic nodes of mesoscopic circuits with an unprecedented accuracy. We first verified that charge quantization is completely destroyed as soon as a ballistic channel connects the metallic island/node to the circuit.

At low temperature we compared our experimental data to the quantitative theory in the two known limits ($\tau_L \ll 1$ and $1-\tau_R \ll 1$ called 'asymmetric'; $1-\tau_{L,R} \ll 1$ called 'symmetric') and we observed a quantitative agreement, within the error bars in the symmetric strong coupling limit. In these two known limits, the visibility on the Coulomb oscillations of the conductance was predicted to scale as $\sqrt{1-\tau_R}$, we actually observed such a scaling for all intermediate transmissions $0 < \tau_L < 1$ in the low temperature quantum regime. This $\sqrt{1-\tau}$ scaling is expected by the theory at high temperatures $k_B T \gg E_C/\pi^2$ as well as the exponential suppression of quantization with temperature that we observed. Moreover, at high temperature, as expected by theory, the degree of charge quantization adopts a fully universal behavior whatever $0 < \tau_{L,R} < 1$, when rescaled with $\sqrt{1-\tau_L}\sqrt{1-\tau_R}$.

References

1. Robert Andrews Millikan, On the elementary electrical charge and the Avogadro constant. Phys. Rev. **2**(2), 109 (1913)
2. T.A. Fulton, G.J. Dolan, Observation of single-electron charging effects in small tunnel junctions. Phys. Rev. Lett. **59**(1), 109 (1987)
3. D.S. Duncan, C. Livermore, R.M. Westervelt, K.D. Maranowski, A.C. Gossard, Direct measurement of the destruction of charge quantization in a single-electron box. Appl. Phys. Lett. **74**(7), 1045–1047 (1999)
4. S. Amasha, I.G. Rau, M. Grobis, R.M. Potok, H. Shtrikman, D. Goldhaber-Gordon, Coulomb blockade in an open quantum dot. Phys. Rev. Lett. **107**(21), 216804 (2011)

5. D.I. Bradley, R.E. George, D. Gunnarsson, R.P. Haley, H. Heikkinen, Y.A. Pashkin, J. Penttilä, J.R. Prance, M. Prunnila, L. Roschier, M. Sarsby, Nanoelectronic primary thermometry below 4 mk. Nature Commun. **7** (2016)
6. Z. Iftikhar, A. Anthore, S. Jezouin, F.D. Parmentier, Y. Jin, A. Cavanna, A. Ouerghi, U. Gennser, F. Pierre, Primary thermometry triad at 6 mK in mesoscopic circuits. Nature Commun. **7**, 12908 (2016)
7. P. Joyez, V. Bouchiat, D. Esteve, C. Urbina, M.H. Devoret, Strong tunneling in the single-electron transistor. Phys. Rev. Lett. **79**(7), 1349 (1997)
8. K.A. Matveev, Coulomb blockade at almost perfect transmission. Phys. Rev. B **51**(3), 1743–1751 (1995)
9. L.P. Kouwenhoven, N.C. van der Vaart, A.T. Johnson, W. Kool, C.J.P.M. Harmans, J.G. Williamson, A.A.M. Staring, C.T. Foxon, Single electron charging effects in semiconductor quantum dots. Z. für Phys. B Condens. Matter **85**(3), 367–373 (1991)
10. N.C. van der Vaart, A.T. Johnson, L.P. Kouwenhoven, D.J. Maas, W. de Jong, M.P. de Ruyter, A. van Steveninck, C.J.P.M. van der Enden, Harmans, C.T. Foxon, Charging effects in quantum dots at high magnetic fields. Phys. B Condens. Matter **189**(1–4), 99–110 (1993)
11. D.B. Chklovskii, B.I. Shklovskii, L.I. Glazman, Electrostatics of edge channels. Phys. Rev. B **46**(7), 4026 (1992)
12. C. Pasquier, U. Meirav, F.I.B. Williams, D.C. Glattli, Y. Jin, B. Etienne, Quantum limitation on Coulomb blockade observed in a 2D electron system. Phys. Rev. Lett. **70**(1), 69 (1993)
13. I.L. Aleiner, L.I. Glazman, Mesoscopic charge quantization. Phys. Rev. B, **57**(16), 9608 (1998)
14. L.I. Glazman, K.A. Matveev, Lifting of the Coulomb blockade of one-electron tunneling by quantum fluctuations. Sov. Phys. JETP **71**, 1031–1037 (1990)
15. C.H. Crouch, C. Livermore, R.M. Westervelt, K.L. Campman, A.C. Gossard, Coulomb oscillations in partially open quantum dots. Superlattices Microstruct. **20**(3), 377–381 (1996)
16. S.M. Cronenwett, S.M. Maurer, S.R. Patel, C.M. Marcus, C.I. Duruöz, J.S. Harris Jr., Mesoscopic Coulomb blockade in one-channel quantum dots. Phys. Rev. Lett. **81**(26), 5904 (1998)
17. S. Jezouin, Z. Iftikhar, A. Anthore, F.D. Parmentier, U. Gennser, A. Cavanna, A. Ouerghi, I.P. Levkivskyi, E. Idrisov, E.V. Sukhorukov, L.I. Glazman, F. Pierre, Controlling charge quantization with quantum fluctuations. Nature **536**(7614), 58–62 (2016)
18. L.P. Kouwenhoven, C.M. Marcus, P.L. McEuen, S. Tarucha, R.M. Westervelt, N.S. Wingreen, *Electron Transport in Quantum Dots*. Kluwer Series, editor, Mesoscopic Electron Transport **E345**, 105–214 (1997)
19. M.H. Devoret, H. Grabert, *Single Charge Tunneling: Coulomb Blockade Phenomena in Nanostructures: [Proceedings of the NATO Advanced Study Institute on Single Charge Tunneling, held March 5–15, 1991, in Les Houches, France]* (Plenum Press, 1992)
20. K.A. Matveev, Quantum fluctuations of the charge of a metal particle under the Coulomb blockade conditions. Sov. Phys. JETP **72**(5), 892–899 (1991)
21. K.A. Matveev, A.V. Andreev, Thermopower of a single-electron transistor in the regime of strong inelastic cotunneling. Phys. Rev. B **66**(4), 045301 (2002)
22. H. Yi, C.L. Kane, Coulomb blockade in a quantum dot coupled strongly to a lead. Phys. Rev. B, **53**(19), 12956 (1996)
23. Y.V. Nazarov, Coulomb blockade without tunnel junctions. Phys. Rev. Lett. **82**(6), 1245 (1999)
24. A. Furusaki, K.A. Matveev, Theory of strong inelastic cotunneling. Phys. Rev. B **52**(23), 16676–16695 (1995)
25. I.P. Levkivskyi, E. Idrisov, E.V. Sukhorukov. Untitled. in preparation
26. K. Le Hur, G. Seelig, Capacitance of a quantum dot from the channel-anisotropic two-channel Kondo model. Phys. Rev. B **65**(16), 165338 (2002)
27. S.E. Korshunov, Coherent and incoherent tunneling in a Josephson junction with a "periodic" dissipation. Sov. J. Exper. Theor. Phys. Lett. **45**, 434 (1987)
28. Gerd Schön and Andrej Dmitievič Zaikin, Quantum coherent effects, phase transitions, and the dissipative dynamics of ultra small tunnel junctions. Phys. Rep. **198**(5–6), 237–412 (1990)

Chapter 3
Observation of the Multi-channel 'charge' Kondo Effect

This chapter is divided in three sections. The first one gives an introduction to the multi-channel Kondo effect, which occurs when a local spin is antiferromagnetically coupled with multiple electron continua. It has become central to study non-Fermi liquid physics, but its experimental observations remained mostly elusive. A powerful implementation called 'charge' Kondo effect will be explained in the second section. The charge model involves 'charge' degrees of freedom instead of 'spin'. The community was doubtful about an experimental realization of this model since it requires apparently contradicting design specifications. This was resolved by implementing the 'charge' Kondo model with a hybrid metal-semiconductor device, and gave us access to the rich multi-channel Kondo physics that will be presented in the last section.

3.1 The Kondo Model: A Testbed for Correlated Physics

The simple Kondo model has attracted a lot of interest because of the associated rich correlated physics. In its original (one-channel) version, it involves a single bath of conduction electrons that tries to screen a magnetic impurity. This model has developed into a testbed for the many-body theoretical methods, in particular the renormalization group. In this picture, as the temperature is lowered, the parameters of the model effectively renormalize to eventually reach universal fixed points, which do not depend on microscopic details.

This renormalization flow takes place on a characteristic temperature called the Kondo temperature T_K. Although at $T \sim T_K$ the one-channel Kondo model involving conduction electrons interacting with a spin-half impurity constitute a many-body problem, at lower temperatures $T \ll T_K$, the system behave as a Fermi liquid because the impurity is screened.

Z. Iftikhar, *Charge Quantization and Kondo Quantum Criticality in Few-Channel Mesoscopic Circuits*, Springer Theses,
https://doi.org/10.1007/978-3-319-94685-6_3

In an attempt to describe real metals, Nozières and Blandin proposed a variant of the original model where the conduction electrons carry multiple 'flavors' that account for additional degrees of freedom. When the number of flavors is larger than twice the spin of the impurity, the impurity is overscreened. This situation results in an antiferromagnetically coupled residual spin on the impurity site; and hence, in a "residual" Kondo effect. As the Kondo impurity is never fully screened, whatever the temperature, a Fermi liquid behavior is never recovered. The multiple channel/flavor Kondo model therefore gives rise to non-Fermi liquid behaviors.

The first part of this section deals with the original Kondo model (where only a single channel is considered) whereas the second part will be about the multi-channel Kondo model.

3.1.1 The Original Kondo Model

This part starts with a description of the model. Then several solutions will be given, following the historical theoretical developments, in order to eventually get a good understanding of the model. We will finally present some experimental realizations in the realm of nano-devices.

The Kondo model and its solution in the perturbative regime

The original model was used by Kondo in 1964 [1] to explain a minimum in the 'resistivity versus temperature' curve of some dilute magnetic alloys (for instance AuFe: gold metal with small amounts of iron impurities). In normal metals, the main contribution to the resistivity is the electron-phonon scattering, a contribution that decreases with the temperature. At zero temperature, the phonon population is zero and the finite residual resistivity is explained by the scattering with the defects of the metal. In contrast, the Kondo model reproduces the actual experimental observations[1] of a logarithmic increase of the resistivity at low temperature due to the scattering by magnetic impurities.

The Hamiltonian of the model is [1]:

$$H_K = \sum_{\vec{k},\sigma} \varepsilon_{\vec{k}} c^{\dagger}_{\vec{k},\sigma} c_{\vec{k},\sigma} + J\vec{S} \cdot \sum_{\vec{k},\vec{k}',\sigma,\sigma'} c^{\dagger}_{\vec{k},\sigma}(\vec{0})\,\vec{s}_{\sigma\sigma'}\, c_{\vec{k}',\sigma'}(\vec{0}) \tag{3.1}$$

The first term corresponds to the kinetic energy of the conduction electrons while the second one describes the scattering with a magnetic impurity of spin $\vec{S}$ localized at the origin $(\vec{0})$. The strength of the interaction between the impurity $\vec{S}$ and the conduction electrons is set by the exchange coupling J. The $c_{\vec{k},\sigma}$ $(c^{\dagger}_{\vec{k},\sigma})$ are operators that annihilate (create) a conduction electron with momentum $\vec{k}$ and spin σ, and $\vec{s}$ is a vector of the Pauli matrices.

[1] The first observations were made in the 30's, for a review see [2].

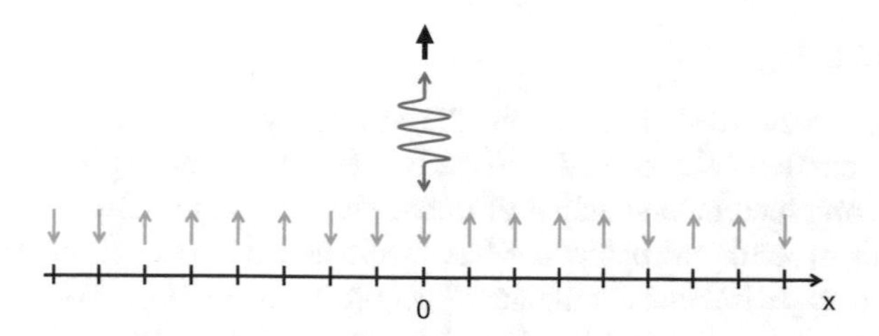

Fig. 3.1 Diagram explaining the Kondo model. The impurity $S = 1/2$ is represented by a red arrow. The conduction electrons (blue arrows) are distributed on a lattice. To simplify, the lattice is 1D. The antiferromagnetic interaction is drawn as a squiggly green arrow located on the impurity site

In the Kondo Hamiltonian, the coupling term acts only on the site of the impurity (located by convention at the origin $\vec{0}$, see Fig. 3.1). This coupling is called 'ferromagnetic' when $J < 0$, in which case the energy H_K is lower when the spin of the electron on the impurity site is parallel to $\vec{S}$. We will rather be interested in the opposite case, $J > 0$, of an 'antiferromagnetic' coupling. In this thesis we will only consider the case of a spin-half impurity, $S = 1/2$. Moreover, while the Eq. (3.1) describes an 'isotropic' Kondo effect, we will also discuss the 'anisotropic' case where the coupling along an axis (say J_z) is different from the other ones ($J_\perp \triangleq J_x = J_y$).

Note also that the impurity states $\vec{S}$ are degenerate; breaking the degeneracy by applying a magnetic field h (say along the z-axis) will generate a term in hS_z in H_K because one of the two projections of $\vec{S}$ will be favored. However, having a degenerate impurity is essential to obtain a fully developed Kondo effect.

Assuming J is small, Kondo used a perturbation theory to calculate the resistivity due to the scattering with the impurity $\vec{S}$ [1]. Extending the calculation to the third order in perturbation led him to the following expression:

$$R(T) = R_0 \left[1 - 2J\rho \log\left(\frac{k_B T}{D}\right)\right] \tag{3.2}$$

where R_0 is a constant, ρ is the density of states and D is a high energy cutoff. This expression was supposed to remain small (as it has been derived using a perturbation theory), but it diverges at $T \sim 0$. Moreover, such a divergence has not been observed experimentally. Actually, Kondo himself was aware that this expression is not valid at low temperature compared to the so-called *Kondo temperature*: $k_B T_K \sim D \exp(-1/J\rho)$.

Although T_K has been initially introduced to delimit the validity of the perturbation theory, it will play an important role in this chapter. Indeed it contains all the microscopic details of the model, which is fully determined by this unique parameter. Results expressed in T/T_K will be universal in the sense that any observable from any experiment should behave identically when plotted with this rescaled temperature.

Poor man's scaling

Actually, Kondo succeeded to explain the intriguing feature of the 'resistivity vs temperature' curves observed in the 30's... but that is just the beginning of the story! Indeed, the complete understanding of the model and, in particular, the problem of what happens around and below T_K has fascinated theorists (even "more than is justified by its experimental significance" according to Wilson [3]).

The key idea is *renormalization*. If we incorporate the log term into J in Eq. (3.2), we could say that the exchange coupling is effectively renormalized[2] to bigger values as the temperature decreases. This concept is actually even more general. Anderson has proposed a *poor man's scaling* consisting in a renormalization of the Hamiltonian to take into account a reduction of the energy bandwidth of the problem [4]. This derivation also is not rigorous, but it allowed Anderson to claim that the coupling J flows towards infinitely large values $J \longrightarrow \infty$ in the antiferromagnetic[3] case.

The fact that J actually flows to infinity or converges to an intermediate fixed point is not obvious. Going to the next order in Anderson's derivation gives an intermediate fixed point, but this is an artifact [5]. The reliable proof that J flows to infinity has been given by Wilson's numerical calculations [3].

The complete solutions

The log divergence in the expression of the resistivity computed by Kondo is a signature of a lack of characteristic energy (Kondo had to sum the contribution of the possible processes on the full range of energies with the same weight). This is typical of some problems of quantum field theory or (classical) critical phenomena where renormalization group theory is of great help.

Wilson has implemented an efficient numerical version of the renormalization and solved the Kondo problem [3]. The key idea of his iterative algorithm is to use a logarithmic discretization to capture all the physics of the problem. Note that this not a perturbative method: the result will be *exact* provided the discretization step is fine enough and the algorithm has converged.

Using this numerical technique, Wilson was able to compute the impurity magnetic susceptibility and its specific heat. But his major result was the demonstration that there is a *unique stable* fixed point in the one-channel Kondo model: $J \longrightarrow \infty$.

Other exact solutions have been found for the Kondo model by the Bethe-ansatz technique [6, 7]. A partial solution consists in a mapping of the Kondo model on the resonant-level model; at a particular value the anisotropic coupling J_z known as the Toulouse point [8], the interaction term of the resonant-level model vanishes and the problem becomes trivial [9] (see also [5]).

---

[2]Everything happens as if the value of J was changing; but note that the true value of the exchange coupling never changes, it is only *effectively* renormalized. We will then distinguish the 'bare' value J_∞ from 'renormalized' value J (both are equal before renormalization).

[3]It flows to zero in the ferromagnetic case, this case can be solved easily [4, 5].

A "Fermi-liquid" description at low temperature

Based on the fact that the antiferromagnetic coupling J to the impurity $\vec{S}$ flows to infinity, Nozières proposed a basic explanation of the physics at low temperature [10]. At zero temperature, the ground state of H_K is simply formed with an electron trapped by the impurity, because the coupling with the impurity is infinite. The remaining electrons are free to visit the other sites of the lattice (except the one of the impurity which remains occupied by a trapped electron).

The impurity and its trapped electron form a rigid singlet complex $S' = 0$. The second term of H_K vanishes when one considers the effective spinless impurity. One eventually gets a Fermi liquid and can, for instance, determine the resistivity or the impurity magnetic susceptibility at low temperature [10].

We have a complete understanding of the one-channel Kondo problem on the theoretical point of view: at low temperatures it is described by the Fermi liquid theory and at high temperatures the perturbation theory holds. The crossover between these two regimes appears around the Kondo temperature $T \sim T_K$. At these intermediate temperatures, the situation is complicated; however, more sophisticated techniques exist (in particular numerical renormalization). Let us now briefly discuss experimental investigations of the Kondo effect at the nano/micrometer scale.

Observation of Kondo effect due to a single localized magnetic impurity

There has been a revival of Kondo effect studies [11] as a result of the development of new experimental techniques. In practice, we are now able to explore the Kondo effect due to a well-characterized *single* impurity, allowing for stringent tests of the theoretical many-body methods.

The most direct example is probably the measurement of a localized magnetic impurity with an STM [12, 13]. In Ref. [12], some cobalt atoms have been deposited on the surface of a clean crystalline gold (111). Since cobalt is magnetic and gold is not, CoAu alloys were known to exhibit Kondo effect. Here the magnetic impurities are at the surface. They are probed with the aid of the STM tip. The microscope is first used to get the topography of the sample and localize the cobalt atoms. When the tip is put close to an impurity, a peak[4] is observed in the differential conductance versus voltage bias. This signature of the Kondo effect appears only in the vicinity of the impurity and at temperature below T_K. In Ref. [12], the Kondo temperature is deduced from a fit of the conductance peak to a model.

Madhavan and co-workers [12] have obtained a Kondo temperature ($T_K \approx 70$ K) much lower than the one measured in bulk CoAu alloys ($T_K > 300$ K). Indeed, the coupling of the impurity to the conduction electrons is weaker in their sample with impurities on the surface. This experiment is a direct observation of the Kondo effect, but one might be frustrated because this is just 'a single picture' as the coupling is fixed. To further understand this effect, one wants 'a movie' of how the Kondo effect develops depending on the coupling strength. For such a program, one needs tunable devices.

---

[4]The shape of the peak is modified by a Coulomb interaction, but this is well understood [12].

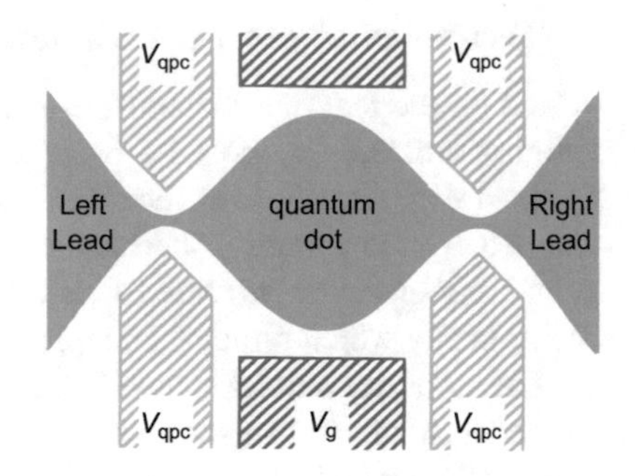

Fig. 3.2 Diagram of a quantum dot connected to two leads. A quantum dot formed in a 2DEG (in solid gray) is connected to two leads through QPCs. The energy in the discrete levels of the dot can be tuned thanks to the gate V_g

Kondo effect in tunable nano-devices

Quantum dots can be fabricated by confining a 2DEG in a small region. Such a 0D structure has discrete levels, and one can populate each level with electrons. A quantum dot can therefore be seen as an artificial atom. The idea in Refs. [14, 15] is to use the two degenerate spin states of a quantum dot (when it contains an odd number of electron) to observe Kondo effect.

In these experiments, a quantum dot connects two leads (like in a Single electron transistor) and the conductance is measured through the dot (see Fig. 3.2). In contrast to the usual Kondo effect (where it is the resistivity which increases at low temperature), here, the conductance increases at lower temperature [11]. This inverted behavior for the quantum dot comes from the fact that, in general, conduction is not allowed through the dot at low temperature because of Coulomb blockade (when no level of the dot is tuned to the Fermi energy). However, in the presence of the Kondo effect, the electrons in the two leads jointly try to make a singlet with the spin of the dot, thus coupling the two leads and increasing the conductance across the quantum dot [14, 15].

The demonstration that this increase in the conductance as the temperature is lowered is indeed due to a Kondo effect is that it happens only when the number of electrons in the dot is an odd number. When there is an even number of electrons on the dot, the total quantum dot spin of the ground state is zero, whereas in the odd case it is $S = \pm 1/2$.

Two years after these pioneer works, in 2000, van der Wiel and co-workers [17] reported an important result (see Fig. 3.3). They measured the conductance versus temperature $G_{\varepsilon_0}(T)$ for several settings of ε_0, the energy of the last occupied level in the quantum dot with respect to the Fermi energy (ε_0 can be tuned thanks to the lateral gate V_{gl}). They then fitted these data to an empirical expression (in red) involving a characteristic temperature called T_K. When plotting these curves using a rescaled temperature $T/T_K(\varepsilon_0)$, they observed that some of the curves collapsed onto a single curve. Thanks to a tunable device, van der Wiel and co-workers were therefore been able to check a major feature of the Kondo effect: its universal behavior.

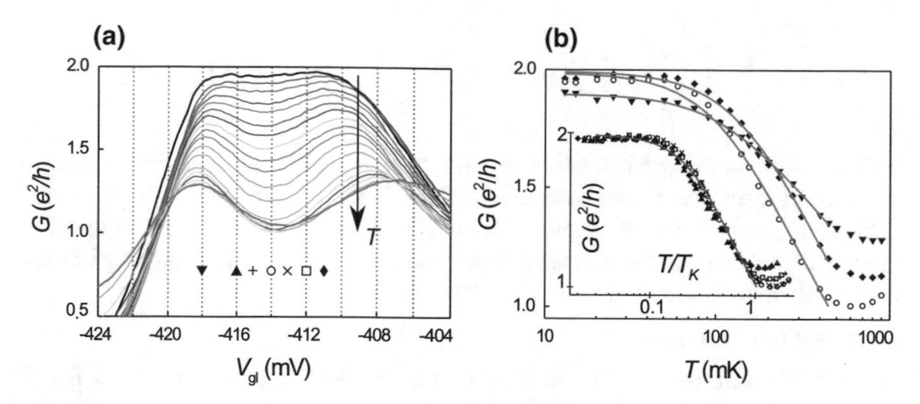

Fig. 3.3 Scaling of the conductance G in T/T_K (reproduced from [16, 17]). **a** Traces of the conductance G versus the lateral gate voltage V_{gl} for temperatures ranging from $T = 15\,\text{mK}$ (top trace) to $T = 900\,\text{mK}$. **b** The conductance is plotted versus temperature for selected values of V_{gl} (the symbols correspond to **a**). Solid lines are fits to empirical law. In the inset, the temperature is rescaled and a universal behavior is observed

3.1.2 The Multi-channel Kondo Model

The original Kondo model shows interesting features, such as renormalization and (consequently) universal behaviors. However, the impurity is perfectly screened at low temperature leading to a trivial situation. But if one considers more electronic channels (or flavors), the impurity may be overscreened giving rise to non-Fermi behaviors.

In this subsection, we will start with the definition of the multi-channel Kondo model. Then we will explain the existence of a fixed point at finite coupling J when the number of channels N is larger than twice the spin of the impurity: $N > 2S$. This fixed point exhibits fascinating non-Fermi liquid behaviors. This will be followed by a short discussion on strongly correlated materials. And we will finally present the first experimental observations of the multi-channel Kondo effect.

The multi-channel model

In order to described the Kondo effect in real metals, Nozières and Blandin have proposed a generalization of the original model called the 'multi-channel Kondo model' [18]. This model involves N channels of conduction, whereas the original model was only considering a single channel. Each channel of conduction is *independent*, just as if the electrons were carrying a flavor (e.g. the orbital degree of freedom). An electron of each flavor can visit each site of the lattice. In particular the impurity site can be occupied by N electrons. As we will soon see, this leads to a completely different qualitative description of the ground state at low temperature. Let me first introduce the Hamiltonian of the multi-channel Kondo model [18]:

$$H_{N-\mathrm{CK}} = \sum_{a=1}^{N} \left\{ \sum_{\vec{k},\sigma} \varepsilon_{\vec{k}} c^{\dagger}_{\vec{k}a\sigma} c_{\vec{k}a\sigma} + J_a \vec{S} \cdot \sum_{\vec{k},\vec{k}',\sigma,\sigma'} c^{\dagger}_{\vec{k}a\sigma}(\vec{0}) \vec{s}_{\sigma\sigma'} c_{\vec{k}'a\sigma'}(\vec{0}) \right\} \tag{3.3}$$

The two models are identical except for the channel index a, in this model. In particular, the original model is obtained if $N = 1$. We can refer to it as '1CK', while 'NCK' will refer to the N-channel Kondo. Notice that each channel is coupled to the impurity S and the channels are not coupled one to another. The coupling of each channel J_a may be different.

Intermediate fixed point

In the 1CK antiferromagnetic model, even the smallest bare value of $J_\infty \neq 0$ will eventually be renormalized to infinity. It means that $J \longrightarrow \infty$ is a stable fixed point while $J = 0$ is unstable.

The position of the fixed points in the multi-channel Kondo model strongly depends on the number of channels. Nozières and Blandin [18] have predicted the existence of a fixed point at finite coupling when the number of channels is larger than twice the value of the spin: $N > 2S$. To illustrate their argument, let us consider our spin-half impurity $S = 1/2$ in the two-channel Kondo model.

First, when $J_{1,2}$ are small, a perturbative approach holds since $T \gg T_K$. The 2CK effect essentially corresponds to twice the 1CK effect. Indeed, as each channel weakly screens the Kondo impurity, both J_1 and J_2 are renormalized independently to larger values.

Second, let us consider the situation where $J_1 = J_2 \triangleq J_{2\mathrm{CK}}$ are infinitely large. In this limit, the impurity traps as many electrons as possible (i.e. the number of trapped electrons equals the number of channels $N = 2$, see Fig. 3.4). The total spin on the impurity site $S' = S - N/2$ is not a singlet in the 2CK case: the impurity is *overscreened*. This situation can be modeled with a new *antiferromagnetic* Kondo effect between S' and the conduction electrons with a weak effective coupling $J'_{2\mathrm{CK}}$. We know that such a fixed point is unstable[5] (just as $J = 0$ was unstable in the 1CK model).

In the 2CK model, both $J_{2\mathrm{CK}} \longrightarrow \infty$ and $J_{2\mathrm{CK}} = 0$ are therefore unstable. Hence, there should be an *intermediate fixed point* towards which the symmetric couplings $J_1 = J_2$ are renormalized as the temperature is lowered. This is in contrast with the 1CK model, where the unique stable fixed point is $J_{1\mathrm{CK}} \longrightarrow \infty$.

Non-Fermi liquid ground state

The ground state of the 2CK intermediate fixed point is qualitatively different from the 1CK fixed point. In the latter case the impurity is completely screened, this yields a non-magnetic impurity which interacts weakly with the conduction electron at $T \ll T_K$. On the other hand, for the 2CK, the impurity is overscreened: The net

[5] In the *underscreened* case $N < 2S$, the residual effective weak coupling to S' is *ferromagnetic*, the fixed point $J_{\mathrm{underscreened}} \longrightarrow \infty$ is therefore stable. This case has been realized in recent experiments [19, 20], however it is not supposed to lead to a non-Fermi liquid ground state (but rather to a 'singular' Fermi liquid one [21]).

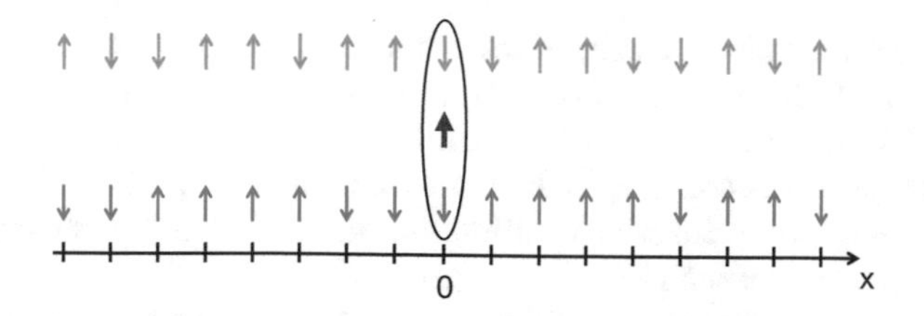

Fig. 3.4 Overscreened impurity in the 2CK model. The impurity traps 2 electrons, one of each channel (blue and orange). This complex (encircled with a black line) has a residual spin, in contrast to the 1CK model. The impurity is *overscreened*, and this situation is unstable

spin on the impurity site never vanishes, and at low temperature, there are still virtual hoppings of electrons of each channel that compete to screen the effective impurity. This nontrivial ground state has some exotic properties; for instance, the impurity has a finite entropy at zero temperature that is equal to $\ln(2)/2$ ([22] and references therein).

The properties of this fixed point have been evaluated using several methods (renormalization group [18, 23] or more recently [24, 25], Bethe-ansatz [26, 27] and CFT [28–30]). As for the original Kondo model, the key quantities to determine are the resistivity, the magnetic susceptibility and the specific heat of the impurity. The exotic properties appear when the number of channel is bigger than twice the impurity spin: $N > 2S$. The 2CK is the simplest model that realizes an overscreened impurity, and hence, it has become a prototype to study non-Fermi liquid behaviors. Indeed both the spin susceptibility of the impurity χ and its specific heat coefficient C_{imp}/T diverge logarithmically (see Bethe-ansatz and conformal field theory techniques references above and also [22, 31]).

Affleck and Ludwig have used CFT techniques to compute the zero-energy amplitude of the single-particle scattering off the impurity [30]. Their expression takes the spin S of the impurity and the number of channels N as parameters. For $S = 1/2$, this amplitude is equal to 1 for $N = 1$, which means that the scattering of a single particle on the Kondo impurity in the $T \longrightarrow 0$ limit simply amounts to a phase shift [32]. In contrast, this amplitude is smaller than 1 for $N > 1$, which means that many body collisions remain important even as $T \longrightarrow 0$ in stark contrast to a Fermi liquid description of free quasiparticles [30, 32, 33].

Emergence of exotic quasiparticles

Majorana fermions are hypothetical spin-half particles predicted by Ettore Majorana to be their own antiparticles. The investigations on Majorana fermions concern fundamental research in both particle physics and condensed matter physics [34]. The motivation in the latter field comes from the proposal [35, 36] that, in some particular geometries, the exchange of Majorana fermions involves topological non-commutative properties that could be used to build a fault-tolerant quantum computer [37, 38].

The low-energy collective excitations at the 2CK quantum critical point can be theoretically modeled with Majoranas. Indeed, using the same idea as Toulouse [8], Emery and Kivelson have shown that at the 2CK fixed point the real part of the electronic excitations decouples, leading to a free Majorana fermion [5, 22, 39] (see also [33, 40, 41]). One can here mention that the 3CK fixed point might involve Fibonacci anyons rather than Majorana fermions.[6]

Let me emphasize that the conditions for these exotic quasiparticles to emerge in the present multi-channel Kondo effect practical implementations require fine-tuning (low temperature $T \ll T_K$, fully degenerate quantum impurity and no channel symmetry perturbation) in contrast to the robust topological properties required for fault-tolerant quantum computation. These emerging particles can probably *not* be used for quantum computation.

Kondo effect in some strongly correlated materials

The ongoing research on some strongly correlated materials and in particular high critical temperature (T_C) superconductor is quite intense. The standard BCS theory sets an upper limit for T_C around 30 K. This limit has been exceeded first in 1986 in a material based on copper oxide [42]. To date, the record is $T_C = 135$ K in normal conditions. The experimental progress raises the hope of practical realization at room temperature. For a recent review on the high-temperature superconductivity in copper oxides, see [43].

Another member in the family of strongly correlated materials are the heavy fermion compounds. In these materials, localized magnetic moments are forming a 'Kondo-lattice'. Heavy fermion compounds can also show non-Fermi liquid physics and superconductivity (up to $T_C \approx 20$ K). The physics in the Kondo-lattices involves a subtle interplay between, on the one hand the Kondo interaction of the conduction electrons with the localized moments, and on the other hand an RKKY interaction between these moments (for a recent review on heavy fermion compounds, see [44]).

Quantum phase transition

What we have discussed so far when dealing with the 2CK overscreened model was the case of two symmetric couplings $J_1 = J_2$. Let us now consider a channel asymmetry.

When we introduce an asymmetry between the bare exchange couplings, say $J_{1\infty} > J_{2\infty}$, the strongest coupled channel (here channel #1) will screen the impurity at low temperature [45] (see Fig. 3.5). The ground state will be similar to the 1CK model: a singlet is formed between an electron of channel #1 and the impurity, and all the other conduction electrons are free. It means that a channel asymmetry in the 2CK model will lead to a Fermi liquid ground state, while it was non-Fermi liquid in the symmetric case.

Any finite initial difference between the bare couplings will diverge under the renormalization process [45, 47]. This critical behavior is a quantum phase transition.

---

[6]Private discussion with A.K. Mitchell; the proposal is from L. Fritz.

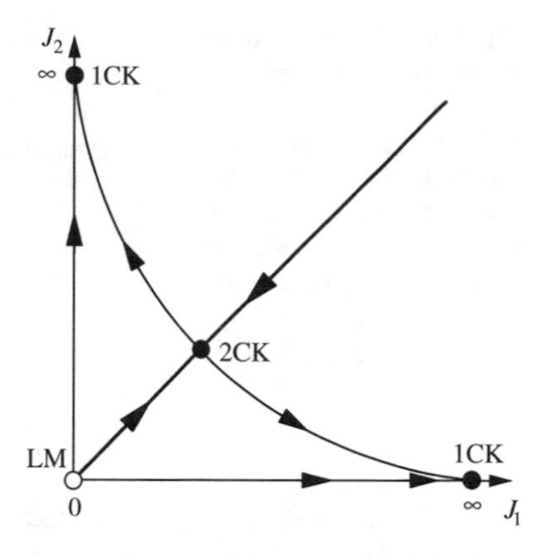

Fig. 3.5 Flow diagram of the 2CK model (reproduced from [46]). The arrows are pointing towards low temperatures. This diagram shows the renormalization of the exchange coupling J_i of each channel. One can see the finite coupling 2CK fixed point and the 1CK fixed points (at infinity). 'LM' is the local-moment fixed point of an uncoupled impurity ($J_1 = J_2 = 0$)

The importance of the 2CK model is partly due to the non-Fermi liquid behaviors and their link with strongly correlated materials and partly due to the richness of the physics of quantum phase transitions. The latter point will be the topic of Chap. 4.

First observation of the two-channel Kondo effect

Despite the fact that the multi-channel Kondo model was originally elaborated to describe the Kondo effect in real metals (with orbital degrees of freedom), no clear evidence of non-Fermi liquid behaviors have been observed in dilute magnetic alloys [45]. The reason for this is probably that there is nothing in these basic systems that forbids a channel asymmetry. A practical application of the multi-channel Kondo model came to the fore in the mid-nineties, with a two-level tunneling system [48].

In this experiment the conductance of a point contact between two pieces of metal have been measured. The point contact acts as a two-level system. The tunneling of electrons through this kind of system have been predicted to show Kondo effect, indeed the spin of the impurity can be mapped onto the two-level system [49].

This implementation can succeed in preserving the channel symmetry, and a non-Fermi liquid power law has been observed for the conductance versus temperature or voltage bias, $G(T, V)$. Moreover, a rescaling of the data shows a universal behavior that was predicted by the theory.

However, the device is not tunable since one can neither change the size of the constriction of the point contact nor the symmetry between the channels. The latter point is important in order to observe a crossover from the non-Fermi liquid to the Fermi liquid physics (quantum phase transition). Moreover, the 2CK interpretation of these results is controversial since the authors of Ref. [50] propose another interpretation [51]. For a recent clarification, based on a 2CK model, see [52].

Multi-channel Kondo effect in tunable nano-devices

We have seen that quantum dots are good systems with which to control the original Kondo effect. To implement the multichannel version, one needs *independent* baths of electron interacting with the quantum dot. This is very challenging, not on a nano-fabrication point of view but rather on the sample design.

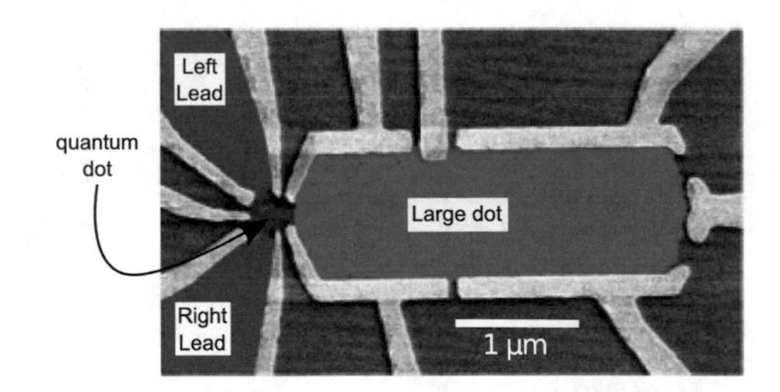

Fig. 3.6 Micrograph of the sample used by Potok and co-workers [53]. The 'Left' and the 'Right' leads (in blue) constitute one channel. The large quantum dot (in red) is another *independent* channel. The quantum dot is at their meeting point (indicated with a green arrow)

In the previous design, two leads were connecting a quantum dot. Each lead is connected to a voltage source and therefore constitutes a bath of electrons. However, since electrons are passing quantum coherently through the dot, there is a connection between the two baths. This does not implement the 2CK but the 1CK model because the two baths of electrons are *not independent*. Oreg and Goldhaber-Gordon have proposed to add a large quantum dot in the original design [54]. The two first leads will constitute the first bath while the large dot will be the second one (see Fig. 3.6). Side gates are tuned in order to forbid charge transfer to the large dot thanks to Coulomb blockade. The two baths are thus independent and this system can be tuned to non-Fermi liquid.

This design has been implemented by the team of D. Goldhaber-Gordon to observe 2CK effect [53, 55]. In the first article, they have measured the non-Fermi liquid power law expected [54, 56]. In the second article they have studied the crossover from non-Fermi to Fermi liquid (when breaking the symmetry between the two channels). Their results are in good agreement with the joined NRG calculations. However, the non-metallic character of the large dot introduces a cutoff for the Kondo scaling physics because of the energy level spacing in this dot (which is finite in practice, but should be negligible in theory). The other main limitation of this observation is the absence of quantitative characterization of the Kondo parameters. In this chapter we demonstrate the two- and three-channel Kondo effect based on an idea proposed by Matveev and Glazman in the early nineties and which allows for the perfect knowledge of the Kondo parameters [57, 58].

3.2 The 'charge' Implementation of the Kondo Model

The Kondo model has different implementations. Indeed, one basically needs a two-level system degenerate in energy, in interaction with N continua. But neither the two-level system need necessarily to be based on spin degrees of freedom, nor the continua need to be of fermionic nature! It then exists a plethora of Kondo effects: even a qubit in interaction with a dissipative environment is a Kondo-type system [59, 60].

In this section, we will consider the ‘charge’ Kondo model, where the degrees of freedom are the charge states of a metallic island weakly coupled to N electrodes with single-mode junctions. Our sample is exactly described by this model which exactly maps onto the multi-channel Kondo model.

We will first explain the analogy between the ‘spin’ and the ‘charge’ Kondo implementation, and then we will give some theoretical predictions for the latter model. At the end of this section, we will discuss the practical implementation of the model.

3.2.1 *Mapping of the Coulomb Blockade Hamiltonian onto the Kondo Model*

In 1991, Matveev [58] drew an analogy between the problem of a metallic island weakly connected to a massive electrode and the multi-channel Kondo model. An exact correspondence between the Hamiltonian of these two problems was established, provided the validity of a few underlying hypotheses.

In this subsection, we will present the ingredients needed to map the two problems by first considering a single channel. Then we will show that it can be naturally extended to the multi-channel model. We will draw the correspondence between the parameters of each model and give a simple picture that explains the analogy. We will finally show that this mapping goes beyond the tunnel limit condition that was used in its derivation.

A piece of metal close to degenerate charge states

Let us first consider a metallic island connected to a massive electrode through a single electronic channel only (see Fig. 3.7). The number of electrons on the island can be changed thanks to a lateral gate V_g (like in a single electron transistor). Close to the degeneracy point $\delta V_g = 0$ between two charge states of the island[7] $Q = 0$ and $Q = e$, the Hamiltonian takes the following form [58]:

$$H = \sum_{k,\alpha} \varepsilon_k c^\dagger_{k\alpha} c_{k\alpha} (\hat{P}_0 + \hat{P}_1) + e\,\delta V_g \hat{P}_1 + t \sum_{k,k'} (c^\dagger_{k1} c_{k'0} \hat{P}_0 + c^\dagger_{k'0} c_{k1} \hat{P}_1) \qquad (3.4)$$

where t is the tunneling probability, α gives the position of the created/annihilated electron: $\alpha = +1/2$ means on the island and $\alpha = -1/2$ means in the electrode. The operators $\hat{P}_\beta$ are projectors on the eigensubspace of the charge operator $\hat{Q}$, which can take two eigenvalues βe with $\beta = 0$ or 1 (see below).

In this Hamiltonian, the first term is the sum of the kinetic energies of the electrons on the island and on the electrode, but limited to the states with $Q = 0$ or $Q = 1$, because we have assumed that we are close to degeneracy $e\delta V_g \ll E_C$ (other charges

[7] Matveev considers a metallic island, which therefore contains a macroscopic number of electrons. The charge Q we are considering is the charge in excess.

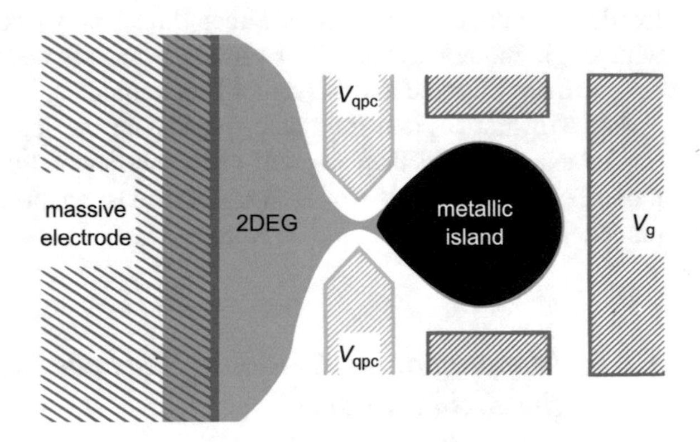

Fig. 3.7 Quantum fluctuation of the charge of a metallic island. A massive electrode (on the left) can exchange single electrons with metallic island (in black) through a single electronic channel made in a 2DEG with a QPC (in the IQHE regime). The plunger gate V_g can be tuned to bring the state with n or $n+1$ electrons on the island to the same energy (the island is then at ‘charge degeneracy’)

states are thus not populated). The second term is the work done by the lateral gate. The last term stands for the tunneling processes: the first term of the parenthesis acts only if $Q = 0$, in which case it annihilates an electron of the electrode and creates one in the island; similarly, the second term transfers a charge e from the island to the electrode. One can easily generalize this Hamiltonian to a version with two electrodes, which will accurately describe a SET close to degeneracy.

Mapping onto the Kondo model

One can note the similarity of the Coulomb blockade Hamiltonian and the one of the Kondo effect Eq. (3.1). Matveev transformed this Hamiltonian and used Pauli matrices to rewrite it in the form of the anisotropic ($J_z \neq J_x = J_y \triangleq J_\perp$) Kondo model:

$$H = \sum_{k,\alpha} \varepsilon_k c^\dagger_{k\alpha} c_{k\alpha} - 2hS^z + \sum_{k,k',\alpha,\alpha'} J_\perp(\sigma^x S^x + \sigma^y S^y)_{\alpha\alpha'} c_{k'\alpha'} c_{k\alpha} \tag{3.5}$$

where h is a magnetic field along the z-axis. Let us emphasize here that the ‘spin’ α involved in this mapping *is not the true spin* σ discussed in the previous Kondo models (the original and the multi-channel). We call this a ‘pseudo-spin’: it can take only two values as it refers to the position of the electron (either on the island or outside). The impurity $\vec{S}$ and location $\vec{\alpha}$ pseudo-spins are thus necessarily equal to $\pm 1/2$.

A simple picture based on pseudo-spins explains this mapping. Indeed, an electron that enters the island flips both the location pseudo-spin $\alpha = \pm 1/2$ and the impurity pseudo-spin $S = \pm 1/2$. This implements the Kondo process quite naively, with an exchange coupling J directly related to the tunneling amplitude t of the Coulomb blockade model.

This implementation of the Kondo effect will be called the 'charge' Kondo model, in contrast to the original (which we be referred as the 'spin' Kondo model). The degrees of freedom involved in each model are of different nature (charge and spin), but both describe the same Kondo effect. The major advantage of this implementation is the natural access it gives to the multi-channel Kondo model as explained below.

A natural implementation of the multi-channel Kondo model

So far we have considered a single-mode junction between the electrode and the island. However, several electronic channels could participate. There are basically three ways to increase the number of channels: (i) to consider the true spin of the electron and say that two identical channels are connecting the island, (ii) to use a wider junction that allows more transverse modes, (iii) to add more electrodes with single-mode junctions.

All these options are modeled identically because electronic channels are assumed independent in the Landauer-Büttiker formalism. One has just to sum over the number N of channels to get the Coulomb blockade or the 'charge' Kondo Hamiltonian:

$$H = \sum_{a=1}^{N} \sum_{k,\alpha} \varepsilon_{ka} c^{\dagger}_{ka\alpha} c_{ka\alpha} (\hat{P}_0 + \hat{P}_1) + e\,\delta V_g \hat{P}_1 + t_a \sum_{k,k'} (c^{\dagger}_{ka1} c_{k'a0} \hat{P}_0 + c^{\dagger}_{k'a0} c_{ka1} \hat{P}_1)$$

$$H = \sum_{a=1}^{N} \sum_{k,\alpha} \varepsilon_{ka} c^{\dagger}_{ka\alpha} c_{ka\alpha} - 2hS^z + \sum_{k,k',\alpha,\alpha'} J_{a\perp} (\sigma^x S^x + \sigma^y S^y)_{\alpha\alpha'} c_{k'a\alpha'} c_{ka\alpha}$$

Thus, there is a correspondence between the parameters of each model. This is summarized in Table 3.1.

The channel symmetry is an important issue in the multi-channel Kondo model. One needs to master the connection of each individual channel to be able to observe the non-Fermi liquid behaviors associated with an overscreened impurity. The option (ii) is thus not suitable. In the option (i) the two (true) spins are identically connected to the island ($t_\uparrow = t_\downarrow$), and the symmetry of the two Kondo channels is thus guaranteed by construction.

However, the option (iii) is more generic. If one can tune manually each electronic channel to get them symmetric, one will naturally access to the N-channel Kondo effect. Moreover, one can break the symmetry between the channels and may observe

Table 3.1 Correspondence between the 'spin' and the 'charge' Kondo models

'spin' Kondo model		'charge' Kondo model	
Impurity spin	$S = \pm 1/2$	Charge state	$Q = 0$ or e
Electrons spin	$\sigma = \{\uparrow, \downarrow\}$	Electron position	{in, out of} the island
Kondo channel	$a = 1, ..., N$	Electronic channel	$1, ..., N$
Exchange coupling	J	Tunneling probability	t
Magnetic field	h	Gate voltage	δV_g

signatures of the expected quantum phase transition. This is actually the option we have used in our experiments.

General validity of the 'charge' Kondo model

One may have noticed two disadvantages of the 'charge' Kondo implementation compared to its original version: (i) the mapping has been made for low tunneling amplitudes t only and (ii) the equivalent exchange coupling is anisotropic by construction ($J_z = 0$ in the 'charge' Kondo Hamiltonian of Eq. 3.5).

Firstly, the validity of the mapping goes beyond the tunnel limit[8] $\tau \ll 1$, as shown in Ref. [61]. In this article, Matveev studied the same problem of the charge quantum fluctuations on a piece of metal coupled to large electrode, but in the opposite limit of an almost perfectly transmitted channel (weak backscattering $1 - \tau \ll 1$). He found a logarithmic divergence of the differential capacitance of the island when approaching the degeneracy $\delta V_g = 0$. Note that the differential capacitance, $C_{\text{diff}} = \partial Q/\partial V_g$, in the 'charge' Kondo model corresponds to the magnetic susceptibility, $\chi = \partial M/\partial h$, of the impurity in the 'spin' Kondo model. His calculations do not require the analogy with the Kondo model, they are rather based on the bosonization technique, and the same logarithmic divergence is found in the tunnel case. This result motivated Matveev [61] to claim that the mapping between the two models is not only valid in the tunnel limit but on the full transmission range $\tau \in [0, 1]$.

This prediction has been demonstrated numerically by Lebanon and co-workers [62]. They have implemented the Coulomb blockade Hamiltonian with a true spin (that plays the role of two symmetric channels) and they have observed a 2CK effect on the capacitance at degeneracy for different values of $t\rho$ (where ρ is a constant density of state) over the full transmission range.

Secondly, the anisotropy of the exchange coupling in the Kondo model is known to be 'irrelevant' for $S = 1/2$ [47]. In the renormalization group language [3, 63, 64], an 'irrelevant' observable is an observable which flows to weaker values under the renormalization process. This means that the results found using an anisotropic model are not less general since this perturbation will reduce under renormalization.

General requirements to implement the 'charge' Kondo model

In the scope of a practical implementation, let me list the requirements to observe the 'charge' Kondo effect:

(i) continuous density of states in the island (negligible level spacing $\delta E \ll k_B T$), which is necessary to map the electrons location onto a pseudo-spin;
(ii) low temperature and voltage bias compared to the charging energy: $k_B T, eV \ll E_C$, necessary to reduce the charge state to a pseudo-spin;
(iii) negligible energy dependence of the transmission $\tau(E)$ on the range explored experimentally $|E| < E_C$, to avoid additional complications (this can effectively lower the energy bandwidth).

---

[8] With our practical implementation in the QHE regime, a single channel is transmitted through the junction below $\tau = 1$. In the tunnel regime, one can identify the transmission with the tunneling probability in this problem $\tau = t^2 \ll 1$.

Table 3.2 Fixed point of the N-channel 'charge' Kondo model

N	$G^*_{N\mathrm{CK}}/G_K$	References
1	0 ?	
2	1	[65, 66]
3	$2\sin^2(\pi/5) \approx 0.691$	[65]
4	1/2	[65, 66]

3.2.2 Theoretical Predictions for the 'charge' Kondo Model

We are interested in theoretical predictions to compare with our experimental observations. The main quantity we will deal with is the conductance G per electronic channel connecting the metallic island.[9] As we have already discussed in Chap. 2, there is a quantitative prediction in the case of two channels almost perfectly connected. This case corresponds to the strong coupling limit of the 'charge' 2CK model.

We need predictions for the two- and three-channel 'charge' Kondo model. The conductance G^* of each of the symmetric channels at the fixed point has been derived in the general N-CK 'charge' Kondo model, with $N \geqslant 2$. As far as I know, there is no prediction for the 1CK, but we will propose a naive guess based on a relation between the exchange coupling J and the transmission τ.

Afterwards we will focus on the power law of the conductance versus temperature near the fixed point, $|G - G^*|(T) \propto (T/T_K)^\gamma$ for the 2- and 3-CK. This scaling depends on the transmission τ of the (symmetric) channels, with $T_K \sim E_C/\delta\tau^{1/\gamma}$, where $\delta\tau = |\tau - \tau_C|$ and τ_C is the bare transmission that corresponds to the fixed point G^*.

The channel conductance and fixed points in the 'charge' Kondo effect

As explained when discussing the multi-channel Kondo model: knowing the position of the fixed point is a crucial and difficult issue. However, Yi and Kane found an expression that gives the conductance G^* at the fixed point of the N-channel 'charge' Kondo model, for all $N \geqslant 2$ [65]!

$$G^*_{N\text{-CK}}/G_K = 2\sin^2\frac{\pi}{2+N} \tag{3.6}$$

This formula agrees with the cases $N = 2$ and $N = 4$ that had already been treated in Ref. [66]. Table 3.2 summarizes the position of the fixed point for the first values of the 'charge' N-channel Kondo model.

The prediction for the 1CK case is not certain, it is based on a argument explained just below.

---

[9] In this chapter, we will consider the individual conductance G of each QPC and not the serial conductance G_{SET} through the whole device as in Chap. 2. For instance, with two symmetric channels $G \triangleq G_1 = G_2$, the serial conductance is $G_{\mathrm{SET}} = G_1G_2/(G_1 + G_2) = G/2$.

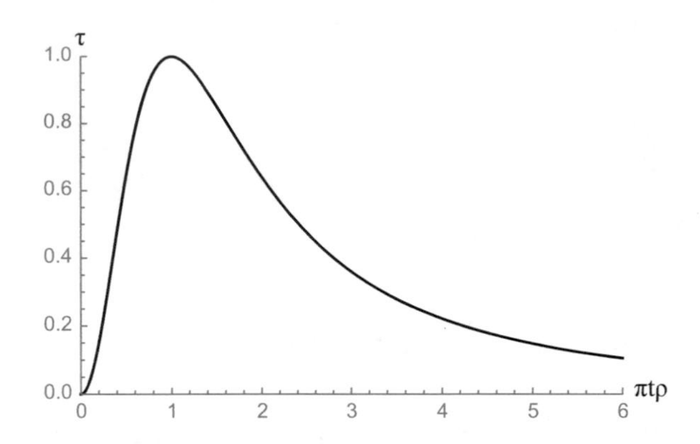

Fig. 3.8 Plot of Eq. (3.7). The transmission τ tends to zero at large t. The maximum $\tau = 1$ is reached at $t = 1/(\pi\rho)$

A naive reasoning to get $G^*_{1\mathrm{CK}}$ using a $J(\tau)$ relation

In the mapping of Matveev [58], the exchange coupling J of the Kondo model is associated with the tunneling matrix element t of the Coulomb blockade model by $J = t$. But the link between the tunneling element t and the transmission τ of a single channel[10] is not monotonous. Hence, the relation between the exchange coupling J and the transmission τ of a channel is non-trivial.

In the absence of charging effect ($E_C = 0$), the transmission τ is a non-monotonous function of ρt (where ρ is the averaged density of charge in the island and in the lead) [62]:

$$\tau = 4\frac{(\pi t\rho)^2}{\left[1+(\pi t\rho)^2\right]^2} \text{ for } E_C = 0 \tag{3.7}$$

A naive reasoning to obtain $G^*_{1\mathrm{CK}}$ is to substitute t by J in this expression, and evaluate it at the 'spin' 1CK fixed point $J \longrightarrow \infty$. The fixed point for the 'charge' 1CK model derived from this naive reasoning is thus $G^*_{1\mathrm{CK}} = 0$.

Prediction for the conductance in the tunnel limit $\tau_{1,2} \ll 1$ in the 2CK model

Furusaki and Matveev have studied analytically the 2CK model. We focus on their prediction for symmetric coupling of the two channels (since the asymmetry between the channels is a relevant perturbation that will drive the system away from the 2CK fixed point [18, 47]).

The two limits treated analytically are those for $\tau_1 = \tau_2 \triangleq \tau \ll 1$ and $1 - \tau \ll 1$. The first one is the tunnel case; the second is the strong coupling case.

In the tunnel case, both channels are first renormalized independently because the couplings are initially weak. The conductance of each channel thus initially increases logarithmically as in the 1CK [66]:

$$G^{\tau\ll 1} = \frac{1}{2}\frac{\pi^2}{\ln^2(T/T_K)} G_K \tag{3.8}$$

[10]In practice realized with a QPC in the QHE regime.

with the scaling Kondo temperature:

$$T_K^{\tau\ll 1} \simeq \frac{E_C}{k_B} \exp\left(\frac{-\pi^2}{2\tau}\right) \tag{3.9}$$

Power law of the conductance versus temperature near the 2CK fixed point

In the strong coupling regime, the quantitative expression of the conductance has been calculated [66]. We already used this expression for the charge quantization in Chap. 2. Here we just focus on the correction proportional to T that vanishes at low temperature at degeneracy (when approaching the 2CK fixed point):

$$G_{2\mathrm{CK}}^{\delta\tau\ll 1} = G_K \times \left(1 - \frac{T}{2T_{K_{2\mathrm{CK}}}^{\delta\tau\ll 1}(\tau)}\right) \tag{3.10}$$

where we have put all the parameters and constants in a coefficient that defines a scaling Kondo temperature in the strong coupling regime $\delta\tau \triangleq |1-\tau| \ll 1$:

$$k_B T_{K_{2\mathrm{CK}}}^{\delta\tau\ll 1}(\tau) = \frac{2E_C}{\pi^3 \gamma \delta\tau} \tag{3.11}$$

All the expressions of scaling temperatures we have given for the 2CK model increase with τ. Indeed, a higher τ set the initial value closer to the fixed point (which is reached at the strong coupling limit $\tau = 1$), the effective temperature $T/T_K(\tau)$ should then be lower.

The strong coupling expression $T_{K_{2\mathrm{CK}}}^{\delta\tau\ll 1}$ diverges at $\tau \sim 1$. Setting τ close to this critical value gives access to Kondo temperatures that are even higher than our cutoff energy E_C! Note that the Kondo model applies even if $T_K > E_C$, since the only practical requirement is $T \ll E_C$.

Power law of the conductance versus temperature near the 3CK fixed point

Using a perturbation theory and a dimensional analysis, Simon and Mora have e-valuated the deviation from the fixed point $\Delta G \triangleq |G - G^*|$ for $N > 2$ number of channels[11] at degeneracy $\delta V_g = 0$ with symmetric couplings τ and at low temperature [67]:

$$\Delta G = c_t \left(\frac{T}{T_K^{\delta\tau\ll 1}}\right)^{\Delta} \tag{3.12}$$

where c_t is a constant of order of 1 and $\Delta \triangleq \frac{2}{2+N}$ is related to the dimension $d_{\mathcal{O}_1} = 1 + \Delta$ of the leading irrelevant operator $\mathcal{O}_1$.

Assuming $T_K^{\delta\tau\ll 1}$ diverges at a critical value τ_C, our collaborators have been able to evaluate $T_K(\tau)$ near τ_C. They have found that the conductance should follow a

[11]The case $N = 2$ is special because the fixed point is reached at an extremal value of $\tau(\rho J) = 1$ (see Fig. 3.8).

Table 3.3 Power laws for the conductance near the N-channel ‘charge’ Kondo fixed point

N	ΔG	$k_B T_K^{\delta\tau \ll 1}$	References
2	$\propto T/T_{K2CK}^{\delta\tau \ll 1}$	$\sim E_C/\delta\tau$	[66]
3	$\propto \left(T/T_{K3CK}^{\delta\tau \ll 1}\right)^{2/5}$	$\sim E_C/\delta\tau^{5/2}$	[67]

scaling law as [29, 68]: $G((T/T_K)^{\Delta}) = G(A\delta\tau(k_B T/E_C)^{\Delta})$ where A is numerical factor of order 1. Hence the scaling Kondo temperature near the fixed point (for all $N > 2$) is:

$$k_B T_K^{\delta\tau \ll 1} \simeq \frac{E_C}{\delta\tau^{1/\Delta}} \tag{3.13}$$

where $\delta\tau \triangleq |\tau - \tau_C|$. We summarize the theoretical laws that we will use in Table 3.3.

Universal scaling

One of the most important features of the original Kondo model is that a unique parameter T_K contains all the microscopic details of the system (energy cutoff D, charging energy E_C, transmission τ). It means that the temperature can be rescaled in T/T_K for any observable to obtain a universal curve. Different experiments realized under different conditions should all collapse on this universal curve after rescaling (whatever the microscopic details).

However, a given observable should renormalize enough to escape a transitory non-universal regime before reaching the universal one ($T \ll E_C, D$). Actually, if one wants to get the *full* universal curve, one should start with a very low T_K (compared to the cutoff energy $T_K \ll D$) [69, 70]. Note that it does not mean that one cannot reach the universal limit starting with a large T_K. The universal limit is actually reached as soon as $T \ll D$ irrespectively of the value of T_K. The only disadvantage of starting with a large $T_K > D$ is that only a *fraction* of the universal curve ($T \ll D < T_K$, close to the fixed point) will be accessible. For instance, if $T_K \sim D$: the upper half of the curve $T/T_K \ll 1$ will be universal because $T \ll D$; whereas the lower half $T/T_K \gg 1$ will be sensitive to the cutoff D and thus non-universal.

Following our publication of a *non-universal* experimental curve of the conductance for the ‘charge’ 2CK model [72], Mitchell and co-workers have modified their powerful interleaved NRG code [24, 73] to implement the ‘charge’ Kondo model. They have been able to start with a small T_K (that corresponds to the tunnel limit, far from the fixed point) and decrease the temperature over many orders of magnitude to get numerically the *full universal* curve of the conductance [70]. This curve is reproduced in Fig. 3.9a. Mitchell has also calculated the universal conductance curve for the 3CK model, but only when approaching the intermediate fixed point from below, see Fig. 3.9b [71].

This universal conductance curve can be compared to the analytical predictions close to the fixed point $T \ll T_K$ (Table 3.3) and also in the tunnel regime $T \gg T_K$ Eq. (3.8) (where the channels are renormalized independently).

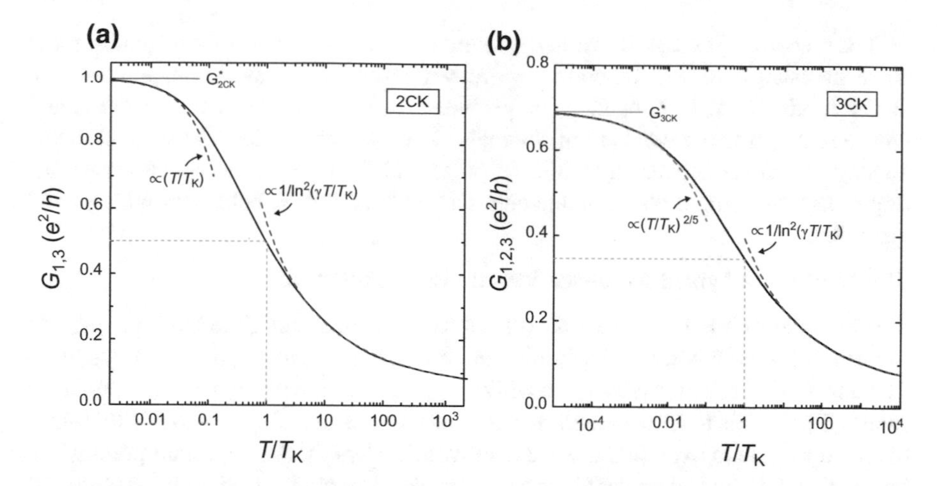

Fig. 3.9 Universal curves of the conductance for the two- and three-channel 'charge' Kondo model [70, 71]. The conductance is plotted versus T/T_K in a log scale. The conductance of the fixed points at indicated with colored solid lines (red for 2CK and green for 3CK). The Kondo temperature T_K is defined such that $G(T = T_K) = G^*/2$ (see grey dashed lines). The colored dashed lines are the analytical prediction in the tunnel regime (blue) and close to the 2CK (red dashed) and 3CK (green dashed) fixed point. These lines are horizontally shifted (γ is a numerical factor) to fit to the universal curve

3.2.3 Experimental Implementation

One may wonder: "But why such hopeful proposal established in the early 90's has never been implemented in practice?". In this subsection, we will see why it is not obvious to design a sample that realizes this model. Then we will explain how we managed to solve this problem.

Contradicting requirements

On page 61, we have listed three necessary hypotheses needed to implement the 'charge' Kondo model. In 2000, Zarárand and co-workers have used a renormalization group approach to evaluate the possibility to observe 2CK behaviors in a SET [74]. In their conclusion, the authors explore two scenarios of experimental implementations to observe 2CK in practice: (i) a metallic island (ii) a 2D island made in a semiconductor.

In the scenario (ii), a good control of the number of channels is possible. One should try to make the smallest possible island to increase E_C because the Kondo mapping only applies at $k_B T \ll E_C$. However, the mean level spacing δE in the 2D island will then not be negligible compared to $k_B T$. But having $\delta E \ll k_B T \ll E_C$ is mandatory to observe the 2CK effect: in the scenario (ii) 2CK will not develop because of the 2D geometry of the island and the small effective mass of the electrons in the 2DEG (for numbers, see Appendix A.2.1).

The scenario (i) does not have any problem concerning density of state requirement since the island is 3D and metallic. Moreover large E_C are available, and one can easily fulfill $\delta E \ll k_B T \ll E_C$. The problem in this case is to connect the island with a few symmetric conduction channels. The authors of Ref. [74] suggest to use an atomic contact between a metallic droplet and the tip of an STM. However this implementation might not be stable enough to observe 2CK behaviors with a good precision.

The solution: a hybrid metal-semiconductor nanostructure

We have been able to conciliate both the requirement of a negligible level spacing and the one of a good control of the number of channel thanks to a hybrid nanostructure. The island of our sample is metallic while the junctions are made in a semiconductor (with QPCs). This nanofabricated structure has all the advantages, and it is therefore the ideal sample to explore multi-channel Kondo physics. An important point when fabricating this hybrid nanostructure is to make a perfect connection between the metallic and the semiconducting part, since otherwise the description of the sample would have to take into account a residual reflection. How well this is achieved and other details about the metallic island are given in Appendix A.2.

Our sample is used in the QHE regime (that breaks the spin degeneracy). It contains three QPCs facing the metallic island; we can then access the two- and three- channel Kondo regimes and observe them through conductance measurements.

3.3 Observation of the Multi-channel 'charge' Kondo Effect

We will now present our experimental observation of the multi-channel Kondo effect using the 'charge' implementation. In particular the experimental data will be compared to the predictions given previously.

This section is divided in two parts, the first one shows the flow of the conductance of the device towards the predicted 2- and 3-CK fixed points (depending on the configuration of the QPCs). The second part focuses on scaling and universality.

The Appendix C explains the experimental procedures to get the data shown in this section while avoiding experimental artifacts. Its reading is not required to understand this section.

3.3.1 Kondo Fixed Points

We will start this subsection by showing Coulomb blockade peaks renormalized by Kondo effect. We will see that the value at the charge degeneracy $\delta V_g = 0$ will tend to the predicted fixed point. These first data already demonstrate that our sample realizes the multi-channel 'charge' Kondo model. Then we will focus on $\delta V_g = 0$ to observe the flow of the conductance towards the fixed point as we lower the temperature.

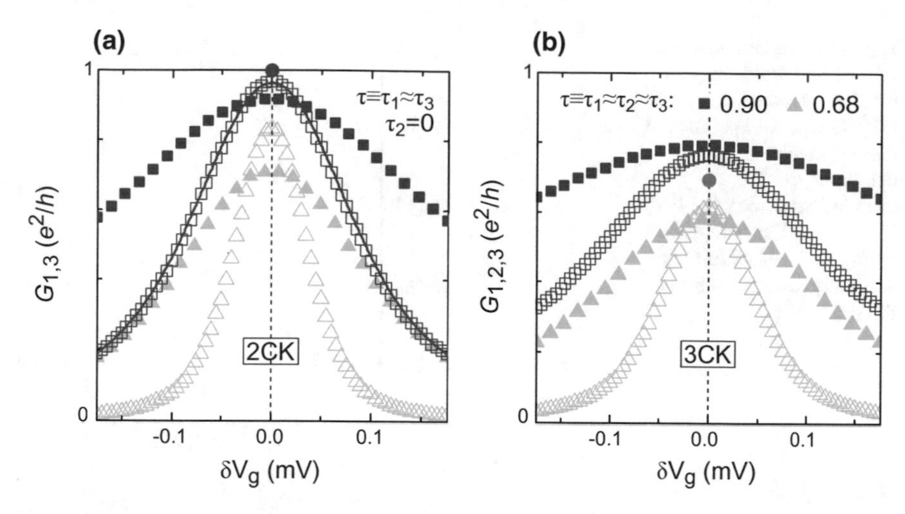

Fig. 3.10 Renormalization of Coulomb blockade conductance peaks by the 2- and 3-channel 'charge' Kondo effect. Coulomb peaks of conductance are plotted as a function of the plunger gate voltage V_g for different temperatures ($T \approx 7.9$ mK for the open symbols and 29 mK for the filled ones) and 2 different symmetric transmissions configurations ($\tau \approx 0.68$ for the triangles and $\tau \approx 0.90$ for the squares) in the 2CK (**a**) and 3CK (**b**) regimes. **a** The conductance $G_{1,3}(\delta V_g = 0)$ flows towards the 'charge' 2CK fixed point $G^*_{2\text{CK}} = G_K$ when lowering the temperature. The red line is the quantitative theoretical prediction of the 'charge' 2CK model plotted for the independently measured parameters. **b** In the 3CK case, the individual symmetric conductances at degeneracy $G_{1,2,3}(\delta V_g = 0)$ flows to the 3CK fixed point $G^*_{3\text{CK}} \approx 0.691\, G_K$

Coulomb peak renormalization by Kondo effect

The Coulomb blockade oscillations of the conductance observed in our sample are renormalized due the 'charge' Kondo effect. The gate voltage V_g acts as an effective magnetic field that destroys the effect, which is thus maximal at charge degeneracy $\delta V_g = 0$ (zero effective magnetic field). Figure 3.10 shows Coulomb peaks in the 2CK and 3CK regimes for two temperatures and two transmissions. As predicted by theory, we observe a renormalization towards the predicted fixed points as the temperature is lowered.

In the 2CK case the conductance at degeneracy increases with the temperature for both transmissions. Away from degeneracy $\delta V_g \neq 0$, we observe however an opposite behavior. This decrease is simply explained by the fact that the Kondo effect is destroyed in this region. The conductance is then described by DCB which blocks the current at low energies (low temperature here). This explains that for a given transmission, the curves at two different temperatures are crossing each other.

The situation is different in the 3CK case where the fixed point is at an intermediate value between 0 and G_K. At $\delta V_g = 0$, when we lower the temperature (passing from filled symbols to open ones), the conductance either increases (triangles) or decreases (squares) depending on the initial position with respect to $G^*_{3\text{CK}}$ (green disk). This signal is quite strong: Below this fixed point (triangle symbols) the traces at two different temperatures are crossing, but above the fixed point (square symbols) they are not.

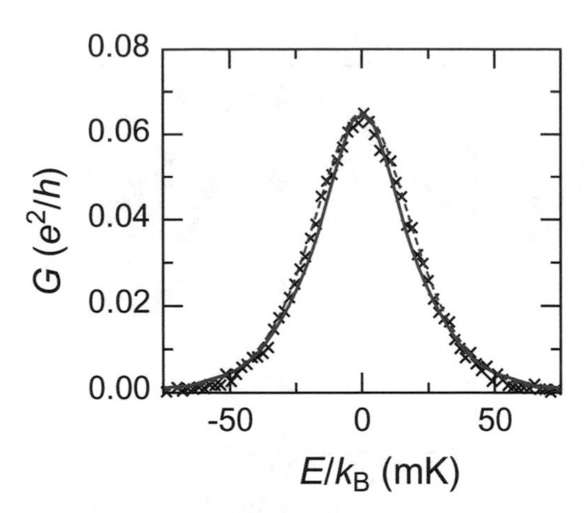

Fig. 3.11 Renormalization of a 2CK tunnel Coulomb blockade conductance peak [70]. Our 2CK experimental data (symbols) [72] are compared to NRG calculations (blue solid line). The standard 'orthodox' theory (red dashed line) is vertically rescaled to have the right height

From a very qualitative point of view, this figure provides a first indication of multi-channel Kondo effect. Moreover, the red line in the 2CK figure shows a plot of the theoretical prediction Eq. (2.9) without any[12] fitting parameter. The transmissions $\tau_1 \approx 0.891$ and $\tau_3 \approx 0.901$ (see Fig. A.6, for the characterization of these values), the temperature $T = 7.90 \pm 0.06$ mK and the charging energy $E_C = 25.8 \pm 0.5\,\mu$eV have all been measured independently to get this quantitative comparison.

Renormalization of the conductance in the 2CK tunnel regime $\tau \ll 1$

Although in the tunnel regime $\tau \ll 1$, the Kondo temperature is very small $T \gg T_K$, the Coulomb peak is subject to a small renormalization that slowly (logarithmically) vanishes at large temperatures (see Eq. 3.8). We have observed [72] that a tunnel peak can be well fitted with the 'orthodox' Coulomb blockade theory Eq. (B.13) with a free amplitude which is in agreement with Eq. (3.8).

The full Kondo renormalization holds only close enough to the degeneracy $\Delta E \leqslant k_B T$ [66], where $\Delta E \triangleq 2E_C \delta V_g / \Delta V_g$ (ΔV_g is the period of the Coulomb blockade oscillations) measures the level splitting between the two charge states of the island (which are degenerate at $\Delta E = 0$). Consequently, a Kondo renormalized tunnel Coulomb peak should be slightly narrower than a usual tunnel Coulomb peak.

This has been verified by the NRG calculation of Mitchell and co-workers. In their figure (reproduced in Fig. 3.11), they have compared their calculations (blue line) to our data (symbols). They have also plotted an 'orthodox' peak renormalized to have the same height (and which is thus broader). We see a slight difference comparable to experimental accuracy.

2- and 3-CK 'charge' Kondo fixed point

Let us now focus on the points where the Kondo effect is not destroyed by energy level splitting: $\delta V_g = 0$. To observe the renormalization towards the fixed points, we

[12]We have only adjusted the position of the maximum of the peak at $\delta V_g = 0$.

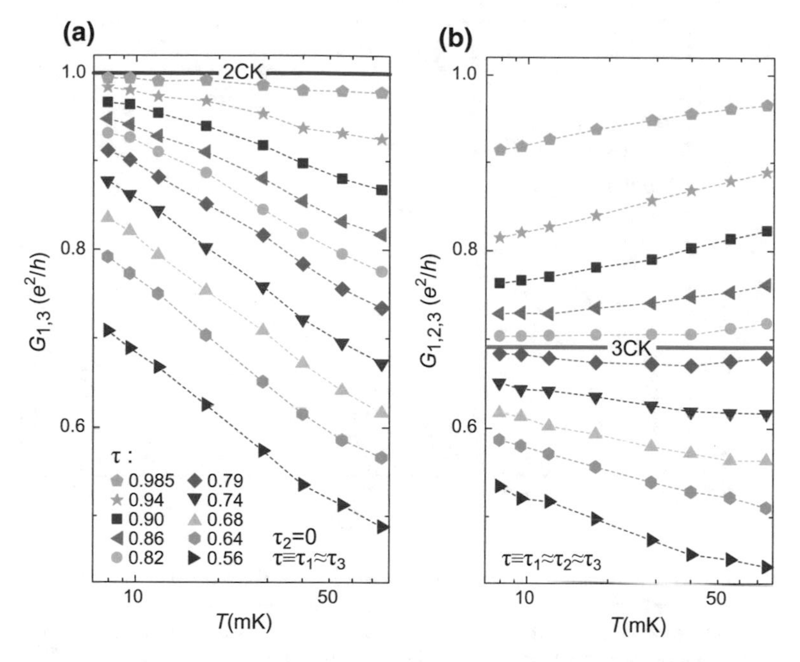

Fig. 3.12 Renormalization towards the two- and three-channel 'charge' Kondo fixed point. For the same set of transmissions τ, the individual conductances at degeneracy $\delta V_g = 0$ are plotted versus the temperature $T \approx \{7.9, 9.5, 12, 18, 29, 40, 55, 75\}$ mK in log-scale. **a** The conductance flows towards the 2CK fixed point (red thick line) for all the transmissions in the case of two symmetric channels ($\tau_1 \approx \tau_3$). **b** With three symmetric channels, it flows to $G^*_{3\mathrm{CK}}$ (green thick line) when lowering the temperature

will look at the temperature dependence of the symmetric individual conductances. This convergence at low temperature is shown in Fig. 3.12 for both 2CK and 3CK.

In the 3CK regime, the conductance flows to a non-trivial universal value $G^*_{3\mathrm{CK}}/G_K = 2\sin^2(\pi/5) = (3-\varphi)/2 \approx 0.691$, where $\varphi = (1+\sqrt{5})/2$ is the golden ratio. The accuracy of this observation is visible in the Fig. 3.13 discussed below. This is the first experimental demonstration of a Kondo coupling strength reaching a universal intermediate fixed point.

3.3.2 *Kondo Scaling*

The experimental data shown in Fig. 3.12 can be compared to two theoretical predictions. The first part of this subsection focuses on the convergence towards the fixed points and checks a scaling to fit to the theoretically predicted power laws. The second one compares the experimental data to the full universal curve of conductance obtained by NRG. In the last part, we discuss the non-universal effects that appear at high temperature.

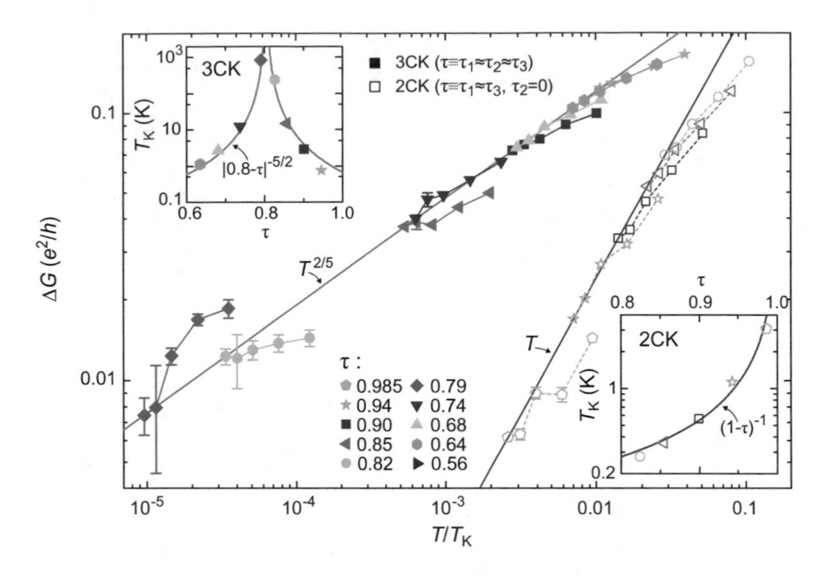

Fig. 3.13 Scaling of the conductance near the two- and three-CK fixed points. The distance $\Delta G \triangleq |G - G^*|$ of the conductance from the fixed point is plotted versus a rescaled temperature. Each transmission corresponds to a symbol (solid for 3CK and open for 2CK). By construction, the lowest temperature point of each set of symbol is on the theoretical law (solid lines). The $T_K^{\delta\tau\ll 1}(\tau)$ used for this scaling are given in the insets and compared with the theory (solid lines)

Power law and scaling in the vicinity of the fixed points

In the vicinity of the Kondo fixed points, the conductance is predicted to scale as a power law $\Delta G \propto (T/T_K^{\delta\tau\ll 1})^\gamma$, where ΔG is the distance between the conductance G and the predicted fixed point G^*, and γ is also given by the theory ($\gamma_{2\mathrm{CK}} = 1$ and $\gamma_{3\mathrm{CK}} = 2/5$). Note that, the scaling Kondo temperature is defined up to a fixed factor (see Table 3.3).

In order to explore this scaling, we plot in Fig. 3.13 the expected power law (solid straight lines), and fix $T_K^{\delta\tau\ll 1}$ so that the experimental point at the lowest temperature ($T \approx 7.9$ mK) matches the theoretical power law. We observe consistency between experimental data and theoretical power laws. We also observe that the data closely approach the predicted fixed point ($\Delta G < 0.01 e^2/h$ for both 2CK and 3CK).

The corresponding Kondo temperatures $T_K^{\delta\tau\ll 1}(\tau)$ are shown as symbols in the insets of Fig. 3.13 and compared with the appropriate theoretical predictions (which are vertically shifted since the $T_K^{\delta\tau\ll 1}$ are defined up to a prefactor). We observe that T_K diverges near a critical value of transmission τ_C. Indeed, one can get arbitrary large scaling Kondo temperatures T_K by setting the transmission such that after an initial Coulomb renormalization, the conductances are closer to the fixed point. For the 2CK case, the predicted critical transmission is $\tau_{C_{2\mathrm{CK}}} = 1$ (there is no Coulomb renormalization for this integer value). In the 3CK case, we find it close to $\tau_{C_{3\mathrm{CK}}} \approx 0.8$

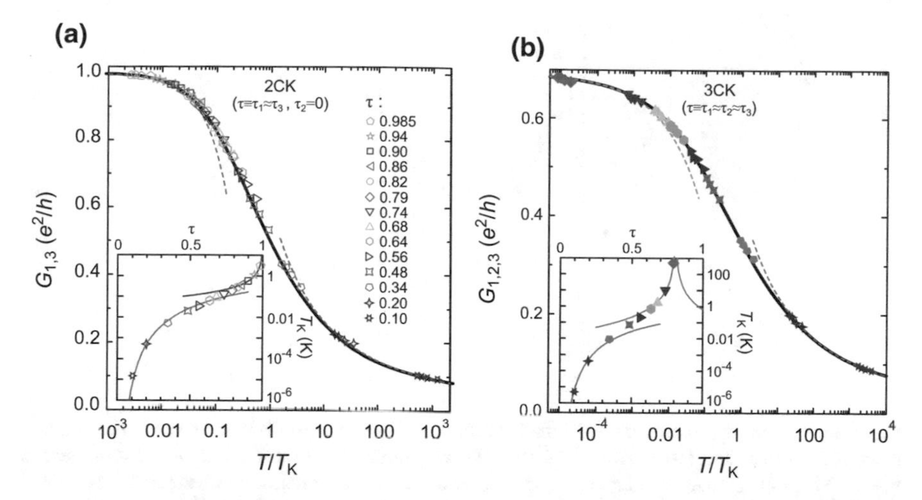

Fig. 3.14 Comparison of the experimental data to the universal curves of the conductance for the two- and three-channel 'charge' Kondo model. Here, experimental data are displayed up to $T = 29\,\text{mK}$ and some additional transmissions are shown compared to Fig. 3.12. Each set point at τ fixed is shifted in the semi-log representation so that the lowest temperature point matches the theoretical curve (solid black line). This defines a scaling temperatures $T_K(\tau)$ that are plotted in the insets for both the 2CK (**a**) and 3CK (**b**) configurations. The lines in each graph are the theoretical prediction shown in Fig. 3.9. In the insets: the blue lines correspond to the theoretical prediction in the tunnel regime; the red and green lines are the predictions for $T_K^{\delta\tau \ll 1}$ close to the 2CK and 3CK fixed points respectively

(which is different from $G^*_{3\text{CK}}/G_K \approx 0.691$ because of the Coulomb interaction[13]). We see that both ΔG and $T_K^{\delta\tau \ll 1}(\tau)$ are in a good agreement with the non-Fermi liquid power laws of the two- and three-channel 'charge' Kondo effect.

Comparison with a universal scaling

The power law of the conductance versus temperature and its scaling in the vicinity of the fixed point are actually contained in the full universal scaling curves displayed in Fig. 3.9. We can directly compare our last experimental data to these exact NRG calculations. In Fig. 3.14, we rescale the temperature in $T/T_K(\tau)$. Each set of data for a fixed τ is adjusted to match the point at the lowest temperature $T \approx 7.9\,\text{mK}$ with the theoretical universal curve. The first remark is that the remaining points ($T > 7.9\,\text{mK}$) are following the universal prediction well. We notice deviations at large temperature, but this is due to the finite E_C of our sample. We find that for 2CK, the three lowest temperatures ($T < 12.0\,\text{mK}$) are in the universal regime (for all τ). The 3CK universal curve is smoother than the 2CK (compare both T/T_K ranges), this may explain why the four lowest temperature data matches the universal 3CK scaling curve.

The scaling Kondo temperature we have used to match the first point on the universal curve also contains information: we can compare it to the theoretical expressions

[13]This initial renormalization occurs at energies of order of E_C.

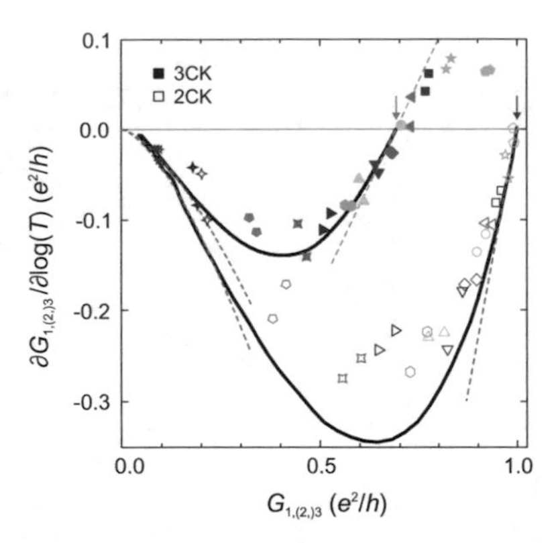

Fig. 3.15 Scaling-free comparison of the experimental data to the universal curves of the conductance for the two- and three-channel 'charge' Kondo model. Finite difference of points shown in Fig. 3.14 is plotted versus conductance (open symbols of two-channel, solid ones for three-channel Kondo). Three temperatures were only considered $T = 7.9$, 12 and 18 mK. Solid curves are the theoretical beta functions, they correspond to the differentiation of the universal NRG conductance curves. Blue dashed lines display the weak-coupling theory and green (3CK) and red (2CK) dashed lines correspond to the NFL power-laws

of $T_K(\tau)$. This is shown in the insets of the figure where one can see that the data near the critical transmission ($\tau_{C_{2CK}} = 1$ and $\tau_{C_{3CK}} \approx 0.8$) and the scaling behavior predicted for $T_K^{\delta\tau \ll 1}(\tau)$ (red line for the 3CK and green line for the 3CK) are in good agreement. The theoretical lines have been vertically shifted to fit to the data at $\delta\tau \ll 1$ to account for the ill-defined prefactor of T_K. We have also plotted $T_K^{\tau \ll 1}$ (in blue solid), to compare with the data close to the tunnel limit $\tau \ll 1$.

Actually one can compare these experimental data to the NRG calculations without scaling. The idea is to consider the finite difference of the experimental points and compare it to the derivative of the NRG curves. The derivative indicates the strength of the renormalization, it is an important quantity known in quantum field theory as the *beta function* of the problem $\beta(G) \triangleq \partial G/\partial \log T$. In Fig. 3.15 we observe a good agreement between the experimental finite differences and the two- and three-channel 'charge' Kondo beta functions. Note that the three-channel beta function has been computed only for $G < G^*_{3\text{CK}} \approx 0.691\, e^2/h$. The experimental points above $G^*_{3\text{CK}}$ therefore constitute to a certain extent a novel quantum simulation that is out of reach of current theoretical methods.

Crossover to universality

In Fig. 3.16a, we show the two-channel Kondo experimental data at higher temperatures. We note that the deviations to the universal curve that develop at high temperatures are all in the same direction. These are not experimental artifacts in the sense that our data are still modeled by the 'charge' Kondo Hamiltonian, but we are sensitive to the finite charging energy E_C. Mitchell and co-workers [70] have reproduced our experimental data with their numerical calculations by taking into account the effect of the cutoff energy D. A more recent work is presented in

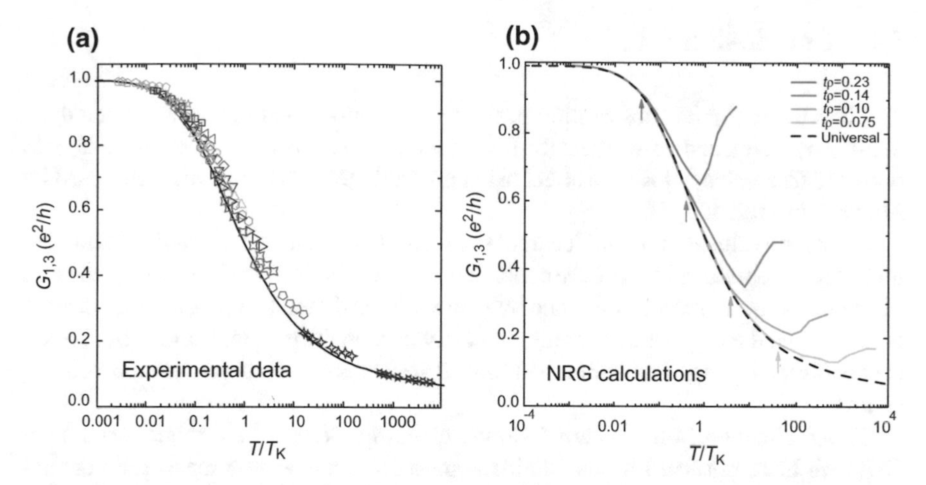

Fig. 3.16 Crossover to the universal two-channel 'charge' Kondo physics as temperature is reduced. **a** Same as Fig. 3.14a, but on a wider range of temperatures: $T =$ 7.9, 9.5, 12, 18, 29, 40, 55 and 75 mK. **b** (A.K. Mitchell calculations [71]) Two-channel Kondo NRG conductance curves calculated for different transmissions. Curves are horizontally shifted (T_K-rescaled) to match the universal one. The arrows indicate the temperature where the deviation to the universal curve gets higher than $0.01e^2/h$. This occurs at $T_{\text{uni}} = E_C/20$ for all transmissions and whatever the cutoff energy considered (here $E_C = 0.1D$)

Fig. 3.16b. The universal curve (dashed black line) is obtained by considering very low transmission, with tunnel amplitude $t = 0.025$. Curves obtained for higher transmissions are T_K-rescaled in a similar way as what we did with experimental data. With the NRG prediction also, we observe a deviation[14] from the universal curve at high temperatures, an arrow indicates a deviation of $0.01e^2/h$ for each curve.

Remarkably, the temperature T_{uni} corresponding to the "less than $0.01e^2/h$ deviation" criterion is independant of T_K/D, T_K/E_C or E_C/D. A.K. Mitchell, who led this numerical study [71] showed that[15] $k_B T_{\text{uni}} \approx E_C/20$. For our device, $T_{\text{uni}} \approx 15$ mK, meaning that the three lowest temperatures are in the universal regime according to our criterion. At higher temperatures, we observe non-universal effects due to the finite value of $E_C \approx k_B \times 300$ mK. The Fig. 3.16 shows the crossover to the universal regime as the temperature is reduced, this is direct observation of the renormalization process that eliminates *irrelevent* perturbations.

[14]The NRG curves also present a minimum, this is a numerical artifact. The minimum occurs at $T = E_C/k_B$ and the part of the curve with $T \geqslant E_C/k_B$ should not be considered.

[15]More generally $k_B T_{\text{uni}} \approx \min(E_C, D)/20$.

3.4 Conclusion

The first unambiguous observation of the two-channel Kondo effect has been done in 2007 by Potok and co-workers [53]. Here we have reported an observation of both two- and three-channel Kondo effect using the 'charge' implementation proposed by Matveev in 1991 [57, 58].

In this implementation, the 'charge' degrees of freedom of a metallic island play the role of a pseudo-spin. We have conciliated the conflicting requirements on the size of the island and its density of state by using a hybrid metal-semiconductor design. Remarkably, the degenerate quantum impurity used in the present Kondo effect study is constituted of large number of electrons: it involves macroscopic quantum charge states!

This realization of the 'charge' Kondo effect implements the model accurately [70]. We have observed a quantitative agreement between the measured conductance and the theoretical prediction without any fitting parameter. We have also compared our experimental data to the universal curve of conductance for both the two- and three-channel Kondo effect. At least our three lowest temperature data, for $T < 15\,\mathrm{mK}$, lie in the universal regime; whereas non-universal behaviors appear at larger T due to the finite charging energy E_C. Hence, when lowering the temperature, we observe a crossover from a non-universal to the universal regime. The comparison between the experiment and the NRG calculations shows that our device fully implement the 'charge' Kondo model, also beyond the universal regime where all the microscopic parameters (τ, E_C, D) can be encapsulated in a single parameter T_K.

The observed non-Fermi liquid power laws for the conductance versus temperature are consistent with the predicted theoretical laws for both two- and three-channel Kondo. Note that it is believed that a localized Majorana quasiparticle emerges at the 2CK fixed point [22, 70]. In this chapter, non-Fermi liquid scalings have been explored by tuning the sample right to the quantum critical points. However, a rich physics is contained in the crossover from criticality (either by breaking the channel symmetry $\Delta\tau \neq 0$ or detuning from the charge degeneracy $\delta V_g \neq 0$), and this will be the topic of the next chapter.

References

1. J. Kondo, Resistance minimum in dilute magnetic alloys. Progress Theoret. Phys. **32**(1), 37–49 (1964)
2. G.J. van den Berg, J. de Nobel, Les propriétés à basses températures des alliages des métaux ≪normaux≫ avec des solutés de transition. Journal de Physique et le Radium **23**(10), 665–671 (1962)
3. K.G. Wilson, The renormalization group: critical phenomena and the Kondo problem. Rev. Mod. Phys. **47**(4), 773–840 (1975)
4. P.W. Anderson, A poor man's derivation of scaling laws for the Kondo problem. J. Phys. C: Solid State Phys. **3**(12), 2436 (1970)

5. T. Giamarchi. *Quantum physics in one dimension* (Clarendon Oxford, 2004)
6. N. Andrei, Diagonalization of the Kondo hamiltonian. Phys. Rev. Lett. **45**(5), 379–382 (1980)
7. P.B. Vigman, Exact solution of sd exchange model at t = 0. JETP Lett. **31**, 364 (1980)
8. G. Toulouse, Exact expression of energy of Kondo Hamiltonian base state for a particular j_z-value. Comptes rendus hebdomadaires des séances de l'Académie des Sciences Serie B **268**(18), 1200 (1969)
9. P.W. Anderson, G. Yuval, D.R. Hamann, Exact results in the Kondo problem. II. Scaling theory, qualitatively correct solution, and some new results on one-dimensional classical statistical models. Phys. Rev. B **1**(11),4464 (1970)
10. P. Nozières, A "Fermi-liquid" description of the Kondo problem at low temperatures. J. Low Temp. Phys. **17**(1–2), 31–42 (1974)
11. L. Kouwenhoven, L. Glazman, Revival of the Kondo effect. Phys. World **14**(1), 33 (2001)
12. V. Madhavan, W. Chen, T. Jamneala, M.F. Crommie, N.S. Wingreen, Tunneling into a single magnetic atom: spectroscopic evidence of the kondo resonance. Science **280**(5363), 567–569 (1998)
13. J. Li, W.-D. Schneider, R. Berndt, B. Delley, Kondo scattering observed at a single magnetic impurity. Phys. Rev. Lett. **80**(13), 2893 (1998)
14. D. Goldhaber-Gordon, H. Shtrikman, D. Mahalu, David Abusch-Magder, U. Meirav, M.A. Kastner, Kondo effect in a single-electron transistor. Nature **391**(6663), 156–159 (1998)
15. S.M. Cronenwett, T.H. Oosterkamp, L.P. Kouwenhoven, A tunable Kondo effect in quantum dots. Science **281**(5376), 540–544 (1998)
16. W.G. van der Wiel, Electron transport and coherence in semiconductor quantum dots and rings. PhD thesis, Technische Universiteit Delft, 2002
17. W.G. van der Wiel, S. De Franceschi, T. Fujisawa, J.M. Elzerman, S. Tarucha, L.P. Kouwenhoven, The Kondo effect in the unitary limit. Science **289**(5487), 2105–2108 (2000)
18. P. Nozières, A. Blandin, Kondo effect in real metals. Journal de Physique **41**(3), 19 (1980)
19. N. Roch, S. Florens, T.A. Costi, W. Wernsdorfer, F. Balestro, Observation of the underscreened Kondo effect in a molecular transistor. Phys. Rev. Lett. **103**(19), 197202 (2009)
20. J.J. Parks, A.R. Champagne, T.A. Costi, W.W. Shum, A.N. Pasupathy, E. Neuscamman, S. Flores-Torres, P.S. Cornaglia, A.A. Aligia, C.A. Balseiro, G.K.-L. Chan, H.D. Abruna, D.C. Ralph, Mechanical control of spin states in spin-1 molecules and the underscreened Kondo effect. Science **328**(5984), 1370–1373 (2010)
21. P. Mehta, N. Andrei, P. Coleman, L. Borda, G. Zarand, Regular and singular Fermi-liquid fixed points in quantum impurity models. Phys. Rev. B **72**(1), 014430 (2005)
22. V.J. Emery, S. Kivelson, Mapping of the two-channel Kondo problem to a resonant-level model. Phys. Rev. B **46**(17), 10812 (1992)
23. D.M. Cragg, P. Lloyd, P. Nozières, On the ground states of some sd exchange Kondo Hamiltonians. J. Phys. C Solid State Phys. **13**(5), 803 (1980)
24. A.K. Mitchell, M.R. Galpin, S. Wilson-Fletcher, D.E. Logan, R. Bulla, Generalized Wilson chain for solving multichannel quantum impurity problems. Phys. Rev. B **89**(12), 121105 (2014)
25. R. Bulla, T.A. Costi, T. Pruschke, Numerical renormalization group method for quantum impurity systems. Rev. Mod. Phys. **80**(2), 395 (2008)
26. N. Andrei, C. Destri, Solution of the multichannel Kondo problem. Phys. Rev. Lett. **52**(5), 364 (1984)
27. A.M. Tsvelick, P.B. Wiegmann, Solution of the n-channel Kondo problem (scaling and integrability). Zeitschrift für Physik B Condensed Matter **54**(3), 201–206 (1984)
28. A.M. Tsvelick, The transport properties of magnetic alloys with multi-channel Kondo impurities. J. Phys. Condens. Matter **2**(12), 2833 (1990)
29. I. Affleck, A.W.W. Ludwig, Critical theory of overscreened Kondo fixed points. Nucl. Phys. B **360**(2), 641–696 (1991)
30. I. Affleck, A.W.W. Ludwig, Exact conformal-field-theory results on the multichannel Kondo effect: Single-fermion Green's function, self-energy, and resistivity. Phys. Rev. B **48**(10), 7297 (1993)

31. A.M. Sengupta, A. Georges, Emery-Kivelson solution of the two-channel Kondo problem. Phys. Rev. B **49**(14), 10020 (1994)
32. L. Borda, L. Fritz, N. Andrei, G. Zaránd, Theory of inelastic scattering from quantum impurities. Phys. Rev. B **75**(23), 235112 (2007)
33. D.L. Cox, A. Zawadowski, Exotic Kondo effects in metals: magnetic ions in a crystalline electric field and tunnelling centres. Adv. Phys. **47**(5), 599–942 (1998)
34. F. Wilczek, Majorana returns. Nat. Phys. **5**(9), 614–618 (2009)
35. N. Read, D. Green, Paired states of fermions in two dimensions with breaking of parity and time-reversal symmetries and the fractional quantum Hall effect. Phys. Rev. B **61**(15), 10267 (2000)
36. A.Y. Kitaev, Unpaired majorana fermions in quantum wires. Phys. Usp. **44**(10S), 131 (2001)
37. C. Nayak, S.H. Simon, A. Stern, M. Freedman, S.D. Sarma, Non-Abelian anyons and topological quantum computation. Rev. Mod. Phys. **80**(3), 1083 (2008)
38. D. Aasen, M. Hell, R.V. Mishmash, A. Higginbotham, J. Danon, M. Leijnse, T.S. Jespersen, J.A. Folk, C.M. Marcus, K. Flensberg, J. Alicea, Milestones toward Majorana-based quantum computing. Phys. Rev. X **6**, 031016 (2016). Aug
39. H.T. Mebrahtu, I.V. Borzenets, H. Zheng, Y.V. Bomze, A.I. Smirnov, S. Florens, H.U. Baranger, G. Finkelstein. Observation of Majorana quantum critical behaviour in a resonant level coupled to a dissipative environment. Nat. Phys. **9**(11), 732–737 (2013)
40. P. Coleman, A.J. Schofield, Simple description of the anisotropic two-channel Kondo problem. Phys. Rev. Lett. **75**(11), 2184 (1995)
41. P. Coleman, L.B. Ioffe, A.M. Tsvelik, Simple formulation of the two-channel Kondo model. Phys. Rev. B **52**(9), 6611 (1995)
42. J.G. Bednorz, K.A. Müller, Possible high-Tc superconductivity in the Ba-La-Cu-O system. Zeitschrift für Physik B Condensed Matter **64**(2), 189–193 (1986)
43. B. Keimer, S.A. Kivelson, M.R. Norman, S. Uchida, J. Zaanen, From quantum matter to high-temperature superconductivity in copper oxides. Nature **518**(7538), 179–186 (2015)
44. P. Gegenwart, Q. Si, F. Steglich, Quantum criticality in heavy-fermion metals. Nat. Phys. **4**(3), 186–197 (2008)
45. M. Fabrizio, A.O. Gogolin, Ph. Nozières, Anderson-Yuval approach to the multichannel Kondo problem. Phys. Rev. B **51**(22), 16088 (1995)
46. M. Vojta, Impurity quantum phase transitions. Phil. Mag. **86**(13–14), 1807–1846 (2006)
47. I. Affleck, A.W.W. Ludwig, H.-B. Pang, D.L. Cox, Relevance of anisotropy in the multichannel Kondo effect: comparison of conformal field theory and numerical renormalization-group results. Phys. Rev. B **45**(14), 7918–7935 (1992)
48. D.C. Ralph, A.W.W. Ludwig, J. von Delft, R.A. Buhrman, 2-channel Kondo scaling in conductance signals from 2 level tunneling systems. Phys. Rev. Lett. **72**(7), 1064 (1994)
49. K. Vladár, A. Zawadowski, Theory of the interaction between electrons and the two-level system in amorphous metals. I. Noncommutative model Hamiltonian and scaling of first order. Phys. Rev. B **28**(3),1564–1581, 1983
50. N.S. Wingreen, B.L. Altshuler, Y. Meir, Comment on "2-channel Kondo scaling in conductance signals from 2-level tunneling systems". Phys. Rev. Lett. **75**(4), 769–769 (1995)
51. D.C. Ralph, A.W.W. Ludwig, J. von Delft, R.A. Buhrman, Ralph, et al., Reply. Phys. Rev. Lett. **75**(4), 770–770 (1995)
52. M. Arnold, T. Langenbruch, J. Kroha, Stable two-channel Kondo fixed point of an SU(3) quantum defect in a metal: renormalization-group analysis and conductance spikes. Phys. Rev. Lett. **99**(18), 186601 (2007)
53. R.M. Potok, I.G. Rau, H. Shtrikman, Y. Oreg, D. Goldhaber-Gordon, Observation of the two-channel Kondo effect. Nature **446**(7132), 167–171 (2007)
54. Y. Oreg, D. Goldhaber-Gordon, Two-channel Kondo effect in a modified single electron transistor. Phys. Rev. Lett. **90**(13) (2003)
55. A.J. Keller, L. Peeters, C.P. Moca, I. Weymann, D. Mahalu, V. Umansky, G. Zaránd, D. Goldhaber-Gordon, Universal Fermi liquid crossover and quantum criticality in a mesoscopic system. Nature **526**, 237–240 (2015)

56. M. Pustilnik, L. Borda, L.I. Glazman, J. von Delft, Quantum phase transition in a two-channel-Kondo quantum dot device. Phys. Rev. B **69**(11), 115316 (2004)
57. L.I. Glazman, K.A. Matveev, Lifting of the Coulomb blockade of one-electron tunneling by quantum fluctuations. Sov. Phys. JETP **71**, 1031–1037 (1990)
58. K.A. Matveev, Quantum fluctuations of the charge of a metal particle under the Coulomb blockade conditions. Sov. Phys. JETP **72**(5), 892–899 (1991)
59. M. Vojta, Quantum phase transitions. Rep. Prog. Phys. **66**(12), 2069 (2003)
60. K. Le Hur, Entanglement entropy, decoherence, and quantum phase transitions of a dissipative two-level system. Ann. Phys. **323**(9), 2208–2240 (2008)
61. K.A. Matveev, Coulomb blockade at almost perfect transmission. Phys. Rev. B **51**(3), 1743–1751 (1995)
62. E. Lebanon, A. Schiller, F.B. Anders, Coulomb blockade in quantum boxes. Phys. Rev. B **68**(4) (2003)
63. L.P. Kadanoff, Critical Phenomena, in *Proceedings of the Enrico Fermi Summer School Course* (Academic, New York, 1971)
64. K.G. Wilson, J. Kogut, The renormalization group and the ε expansion. Phys. Rep. **12**(2), 75–199 (1974)
65. H. Yi, C.L. Kane, Quantum Brownian motion in a periodic potential and the multichannel Kondo problem. Phys. Rev. B **57**(10), R5579–R5582 (1998)
66. A. Furusaki, K.A. Matveev, Theory of strong inelastic cotunneling. Phys. Rev. B **52**(23), 16676–16695 (1995)
67. P. Simon, C. Mora, Scaling analysis near the non-Fermi liquid multichannel kondo fixed. Unpublished work
68. C. Mora, K. Le Hur, Probing dynamics of Majorana fermions in quantum impurity systems. Phys. Rev. B **88**(24), 241302 (2013)
69. P. Nozières, Kondo effect for spin 1/2 impurity a minimal effort scaling approach. Journal de Physique **39**(10), 8 (1978)
70. A.K. Mitchell, L.A. Landau, L. Fritz, E. Sela, Universality and scaling in a charge two-channel Kondo device. Phys. Rev. Lett. **116**(15) (2016)
71. Z. Iftikhar, A. Anthore, A.K. Mitchell, F.D. Parmentier, U. Gennser, A. Ouerghi, A. Cavanna, C. Mora, P. Simon, F. Pierre, Tunable quantum criticality and super-ballistic transport in a 'charge' Kondo circuit. 08 (2017)
72. Z. Iftikhar, S. Jezouin, A. Anthore, U. Gennser, F.D. Parmentier, A. Cavanna, F. Pierre, Two-channel Kondo effect and renormalization flow with macroscopic quantum charge states. Nature **526**, 233–236 (2015)
73. K.M. Stadler, A.K. Mitchell, J. von Delft, A. Weichselbaum, Interleaved numerical renormalization group as an efficient multiband impurity solver. Phys. Rev. B **93**(23), 235101 (2016)
74. G. Zaránd, G.T. Zimányi, F. Wilhelm, Two-channel versus infinite-channel Kondo models for the single-electron transistor. Phys. Rev. B **62**(12), 8137 (2000)

Chapter 4
Quantum Phase Transition in Multi-channel Kondo Systems

Quantum criticality accounts for the unconventional physics that develops in the vicinity of the critical point of a second-order quantum phase transition [1]. It is characterized by the power law divergence of the correlation length ξ as a non-thermal parameter g approaches a critical value g_c: $\xi \propto |g - g_c|^{-\nu}$, where ν is called a critical exponent. The concept of quantum criticality provides a powerful universal framework to describe some of the most fascinating strongly correlated electrons phenomena, including heavy fermions [2] or high-T_C superconductivity [3].

Although tunable nanostructures would provide ideal systems for the quantitative experimental exploration of quantum criticality, only few examples exhibit clear signatures of second-order QPTs [4–6]. In this chapter, we will show that the 'charge' implementation of the multi-channel Kondo model provides an outstanding testbed for the quantum critical physics.

We will begin with an introduction to quantum criticality. Then we present the theoretical predictions specific to the multi-channel Kondo effect. And finally we show our experimental observation of quantum criticality in our device using in turn the two relevant perturbations we have at our disposal (the gate voltage and the channel asymmetry).

4.1 Quantum Phase Transition

The physics associated with the transition between stable phases of matter has been extensively studied both theoretically and experimentally. The description of second-order phase transitions require powerful many-body techniques among which the *renormalization group* has a central place. Rather than being a pure mathematical trick, this technique has a profound physical consequence, namely *universality*.

Z. Iftikhar, *Charge Quantization and Kondo Quantum Criticality in Few-Channel Mesoscopic Circuits*, Springer Theses,
https://doi.org/10.1007/978-3-319-94685-6_4

This field of research concerns systems as various as black holes and strongly correlated materials [1].

In this section, we will first explain what a quantum phase transition is. Then we will focus on the second-order quantum phase transition and address the notion of *quantum criticality*. We will discuss an observation of quantum criticality in a tunable device based on a carbon nanotube connected to dissipative leads [4, 5].

4.1.1 What Is a Quantum Phase Transition

Definition and classification

In contrast to classical phase transitions that occur at finite temperature, and where only thermal fluctuations have to be considered; quantum phase transitions are driven by a non-thermal parameter g and they occur at zero-temperature $T = 0$ where only quantum fluctuations exist.

A quantum[1] phase transition is characterized by the dependence on g of various quantities, such as the correlation length $\xi(g)$, the magnetic susceptibility $\chi(g)$, the specific heat $C(g)$, or even dynamical quantities as the equilibration time $\tau(g)$.

For a first-order phase transition, some of these quantities are discontinuous. At finite temperature, there will be no critical phenomena, and the system will simply consist of a thermodynamical mixture of phases [7].

A second-order (continuous) quantum phase transition occurs at a quantum critical point $g = g_c$. In the vicinity of this point, the correlation length diverges as a power law: $\xi \propto |g - g_c|^{-\nu}$, where ν is called a critical exponent [8]. Note that for a special type of continuous transition called infinite-order transitions, the dependence on g is exponential. This is the case for instance of Kosterlitz-Thouless transitions.

In this thesis, we will only discuss the quantum phase transition that present quantum criticality at finite temperature, and those are the second-order quantum phase transitions. Because of the divergence of the correlation length at the critical point, the system becomes scale-invariant. The quantum critical point is thus suitable for some theoretical approaches such as the renormalization group or the conformal field theory [9].

Renormalization group theory, universality and relevant perturbations

At the critical point, fluctuations of all wavelength have to be considered with the same weight [10]. The renormalization group theory provides a mathematical tool to treat the critical phenomena that occur in the vicinity of continuous phase

[1]The classification that follows is the same for classical phase transitions, one just has to replace the non-thermal parameter g by T.

transitions. This theory solves the problem iteratively. At each step, the couplings $\{K\}$ are rescaled through a transformation $\mathcal{R}$ [9]:

$$\{K'\} = \mathcal{R}(\{K\}) \tag{4.1}$$

In the space of all couplings' configurations, it may exist a *fixed point* $\{K^*\}$ such that $\{K^*\} = \mathcal{R}(\{K^*\})$. This fixed point is not necessarily a quantum critical point.[2] Actually, the renormalization group tells nothing about the nature of the phases. Yet it describes accurately the physics near the fixed points of critical phenomena.

The same *universal* power laws in t^γ (with $t \triangleq |T - T_C|/T_C$ for classical transition and $t \triangleq |g - g_c|/g_c$ for quantum phase transitions) can be observed for various systems [8]. Different observables (specific heat C, magnetic susceptibility χ, relaxation time τ etc.) lead to different universal critical exponents α, γ, z etc. But few of these critical exponents are actually independent since they can be related using scaling relations that can be derived from the renormalization group [8, 9].

The fact that the same power law dependence were found for very different systems is actually well explained by the renormalization group theory. When approaching the critical point, the 'irrelevant' perturbations that depend on the microscopic details of the system vanish and the behavior becomes universal [10]. Hence many systems can be classified in a given class of universality that will be described with the same laws whatever their exact chemical composition. A class of universality can actually be defined as the basin of attraction of a fixed-point Hamiltonian [11].

4.1.2 Quantum Criticality

As mentioned above, a second-order quantum phase transition occurs at zero temperature precisely at the quantum critical point. In practice it is unreal to reach $T = 0\,\mathrm{K}$ and observe a true quantum phase transition. However, quantum criticality extends on a range of parameter that widens with temperature, up to a high temperature cutoff that depends on the underlying microscopic mechanisms.

In this subsection, we will illustrate in a phase diagram the typical parameter growing range of quantum criticality with temperature. Then we will compare this diagram with an experimental observation in a strongly correlated material. Afterwards, we will distinguish between bulk and impurity phase transitions. We will finally depict an observation of a continuous impurity quantum phase transition studied in a tunable device.

Typical phase diagram of a second-order quantum phase transition

Let us call g the non-thermal parameter that drives a continuous quantum phase transition. This transition occurs at the quantum critical point $g = g_c$, at zero temperature. In the vicinity of g_c and at low temperature, the small thermal fluctuations

[2] For instance, the $J \longrightarrow \infty$ fixed point in the 1CK model is not a quantum critical point.

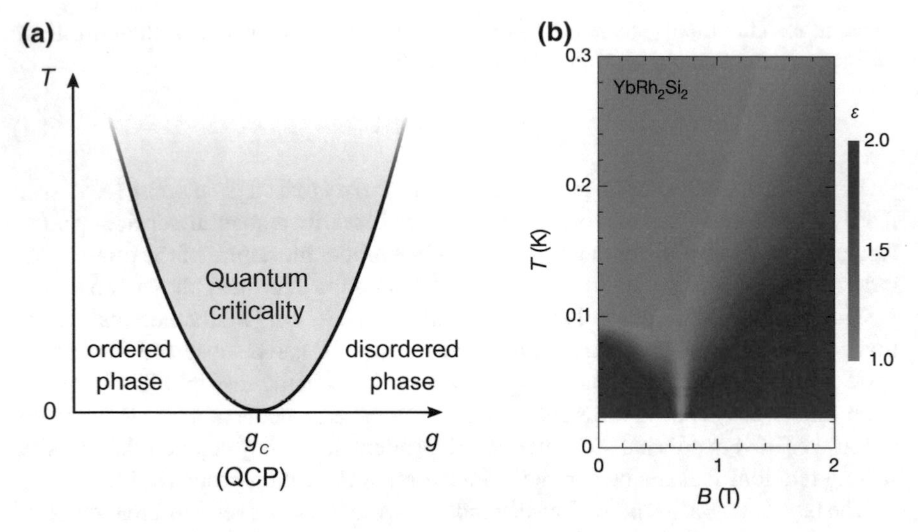

Fig. 4.1 Quantum criticality. **a** Phase diagram near a quantum critical point $g = g_c$. The parameter space for quantum criticality (in orange) widens with temperature. **b** Quantum critical behavior driven by a magnetic field B observed in a heavy-fermion compound (reproduced from [13]). The critical exponent ε of the resistivity versus temperature is non-Fermi liquid ($\varepsilon \neq 2$) in the critical region (in orange)

blur the small difference between $g < g_c$ and $g > g_c$. The system then behaves as if $g = g_c$ and exhibits quantum criticality on a finite range of g [1]. Moreover, the undetermined region widens as the temperature increases! This broadening is shown in Fig. 4.1a. In general, as for usual quantum effects, the quantum critical behaviors vanish at high temperature.[3]

Quantum critical state

The state of matter in the quantum critical region is very different from the two stable phases of the system. Above, we mentioned that the correlation length ξ was diverging at the critical point of a classical phase transition, and the system was then subject to fluctuations of all wavelengths. In the quantum critical state, the system is subject to both quantum and thermal fluctuations. In electronic systems, when these fluctuations diverge, they can give rise to a new type of electronic fluid of strongly correlated electrons with unconventional, non-Fermi liquid behaviors [8].

This singular state of matter is of great interest for the physicists of different fields. There is indeed a link between the conformal field theory used to describe some critical points and string theory ([1, 3] and references therein). For instance, the border of a critical region can be seen as the horizon of a black hole ([1, 14] and

[3]Some strongly correlated materials have displayed a signature of quantum criticality up to 700 K [12].

references therein). Quantum criticality is an active field of research, motivated in particular by the extensive work on strongly correlated materials.

Quantum criticality in strongly correlated materials

Perturbative approaches starting from free electrons fail to account for strongly correlated materials. Generally, strong correlations e.g. in the context of quantum criticality result in a breakdown of the standard Fermi liquid theory description of electronic systems.

This breakdown can be observed as non-Fermi liquid power laws for instance in the resistivity versus temperature. In standard metals, the Fermi liquid theory predicts a T^2 dependence on resistivity ρ. One can plot the critical exponent $\varepsilon = \mathrm{d}\ln\rho/\mathrm{d}\ln T$ versus the non-thermal parameter g to observe a pattern that widens with temperature as shown in Fig. 4.1b. Signatures of quantum criticality have been observed in heavy-fermion [15] and also in other strongly correlated materials (see [1] for examples).

In particular, a T-linear power law of the resistivity versus temperature is observed in the 'strange metal' phase of high-T_C superconductors based on copper oxide [3]. Intriguingly, this critical behavior that originates from a putative quantum critical point located in the superconducting phase can span to the highest attainable temperatures [12]. To date, the theoretical description of this 'strange metal' phase is still essentially incomplete [3].

However, reaching a quantitative microscopic understanding of the superconductivity that emerges from this 'strange metal' phase at low temperature has great practical interest because of its possible technological applications. A first step is to look for systems that exhibit quantum criticality and that are easier to model and to study than the complex, real-world strongly correlated materials. A possibility is to use quantum simulators such as cold atoms in optical lattices [16] or superconducting qubits [17] to emulate bulk quantum phase transitions. An alternative route is to study 'impurity' quantum phase transitions.

'Impurity' quantum phase transition

One can design nano-devices where some local quantum states (the quantum impurity), such as the electronic levels in quantum dots or the charge states of our metallic island, are coupled with thermodynamic baths. When a quantum phase transition occurs in such systems, the only observable that display quantum critical behaviors are those involving the quantum impurity [7].

These systems are easier to model and to treat than 'bulk' systems because of their lower dimensionality. In addition to possibly contributing to a deeper understanding of quantum criticality, 'impurity' problems can also be mapped onto higher dimensionality lattice problems. This technique is used in the dynamical mean-field theory which can treat exactly strongly correlated systems, but only in the infinite dimension limit [18].

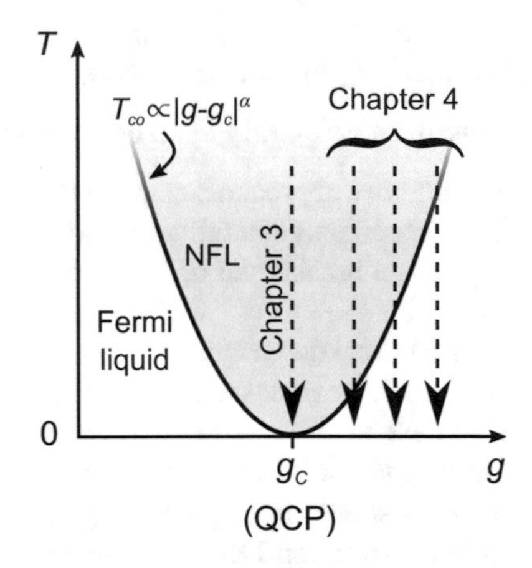

Fig. 4.2 Crossover from quantum criticality. In Chap. 3, we have set the non-thermal perturbations $g = g_c$ and lowered the temperature to observe the power laws when flowing into quantum criticality. In the present chapter, we consider the crossover from the non-Fermi liquid (NFL) quantum critical region (in gray) to Fermi liquid. This crossover occurs on a temperature scale $T_{\text{co}} \propto |g - g_c|^\alpha$

Crossover from criticality

There are basically two ways to explore a quantum critical point. The first one has been the subject of Chapter 3, it to set $g = g_c$ and to lower the temperature T. The second one is to explore the crossover from the critical state to a stable phase (see Fig. 4.2). The power laws associated with these two experiments are sometimes related [8].

Continuous quantum phase transitions in tunable devices

As for the Kondo effect, tunable nano-devices could provide powerful tools for the experimental investigations of quantum criticality. However, observations of quantum criticality in tunable devices are rare. Apart from the implementation of the 2CK model using quantum dot that we have already discussed [6] in Chap. 3, we can mention a realization by Mebrahtu and co-workers [4, 5].

They have studied the development of a second-order quantum phase transition using a carbon nanotube which acts as a quantum dot. The carbon nanotube was connected to two dissipative leads through tunnel junctions. At low temperature, the conductance drops because of dynamical Coulomb blockade, except at the QCP corresponding to symmetric contacts, where the conductance tends to the unitary limit $G_K = e^2/h$. With this device, a non-Fermi liquid power law has been observed on the conductance versus temperature when approaching the quantum critical point, and the broadening of the quantum critical area with temperature has also been studied (see Fig. 4.3).

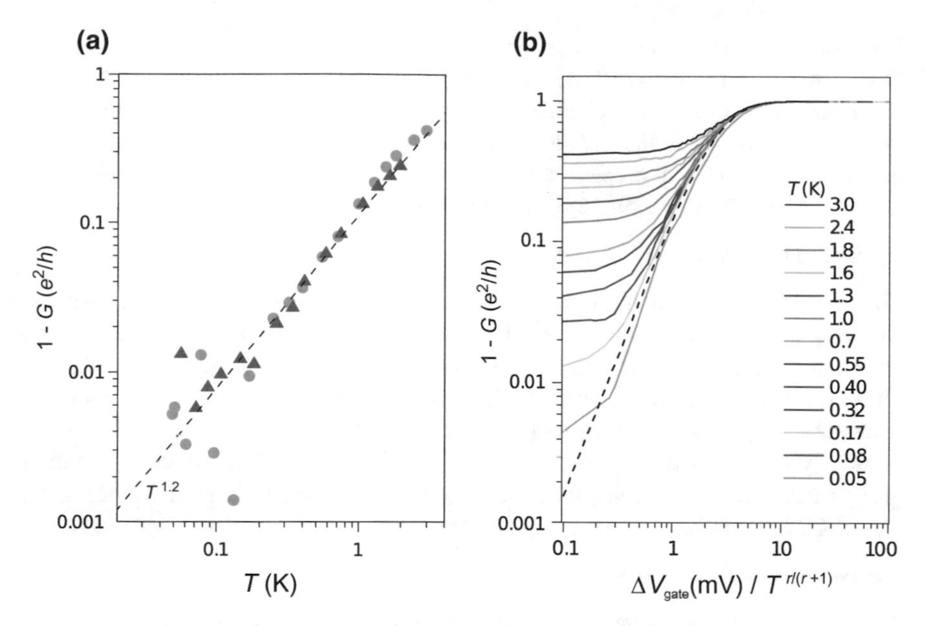

Fig. 4.3 Quantum criticality observed by Mebrahtu and co-workers [5]. **a** A power law $|1 - G| \propto T^{1.2}$ is observed when approaching the quantum critical point. **b** Crossover from criticality when breaking the symmetry using ΔV_{gate}: the larger the temperature, the wider the range where the conductance G saturates close to the unitary limit

4.2 Quantum Phase Transitions in the Multi-channel Kondo Model

A flow towards a non-Fermi liquid fixed point is predicted in the overscreened multi-channel Kondo effect. However, any finite magnetic field or an asymmetry on the channels' bare couplings lead to a different fixed point at low temperature. As detailed below, this quantum phase transition is of second-order and displays non-Fermi liquid quantum critical physics.

In this section we will first consider the general multi-channel Kondo model, discuss quantum criticality and define the crossover temperature. We will also present the crossover from criticality observed by Keller and co-workers in their 'spin' 2CK implementation. Then we will focus on the 'charge' implementation and present the predictions for the two possible relevant perturbations (channel or pseudo-spin asymmetry).

4.2.1 Description of the Quantum Phase Transition

The 2CK model has been extensively studied theoretically and exact results have been obtained. The ground state of this model involves non-Fermi liquid physics.

However, the tiniest (non-zero) channel asymmetry will drive the system towards a Fermi liquid. The crossover from the strongly correlated physics to the free fermion description occurs below a characteristic crossover temperature T_{co} (provided the system was initially in the quantum critical regime $T_{co} < T \ll T_K$). At the end of this subsection, we will present the results of Keller and co-workers who observed this crossover when breaking the channel symmetry [6].

Quantum critical point

The non-Fermi liquid quantum critical physics of the overscreened Kondo impurity is well-established theoretically. Critical exponents of this quantum critical point have been first derived via the Bethe ansatz method [19, 20]. The CFT can be applied to the conformally invariant critical points of the multi-channel Kondo model to get exact results for the power law of the magnetic susceptibility $\chi(T)$, specific heat $C(T)$ or resistivity $\rho(T)$ versus temperature [21, 22]. Of course, numerical renormalization group can also be used to study the fixed points [23]. Because of its relative simplicity, the 2CK model has become a paradigmatic example for the non-Fermi liquid physics.

Crossover from criticality

The asymmetry between the antiferromagnetic couplings (e.g. $J_1 \neq J_2$) is a 'relevant' perturbation in the renormalization group sense (i.e. it grows under renormalization). This is illustrated in Fig. 4.4, which shows that the 2CK intermediate fixed point is unstable under a channel perturbation. This perturbation will drive the system towards a Fermi liquid at low temperature. The smaller the asymmetry ΔJ, the lower the crossover temperature to Fermi liquid will be.

Moreover, one can destroy the Kondo effect by applying a magnetic field that will favor one of the two position of the spin $\vec{S}$ of the impurity. This will drive the system towards a Fermi liquid phase (even for symmetric channels). In the 'charge' implementation, a detuning of the voltage gate δV_g corresponds to an effective magnetic field. In this chapter, we will use the energy level splitting $\Delta E \triangleq 2E_C\delta V_g/\Delta V_g$ rather than δV_g.

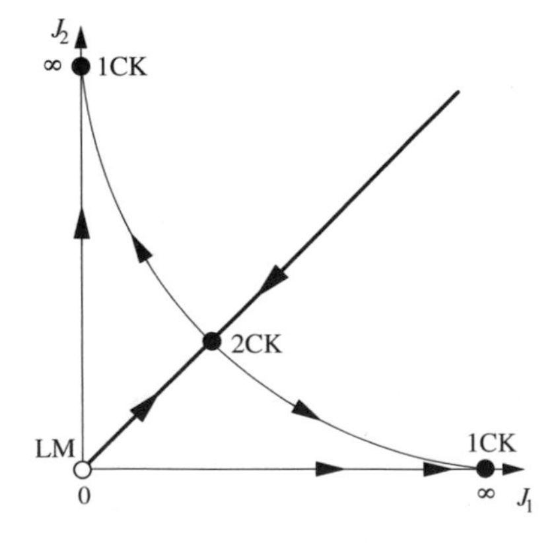

Fig. 4.4 Renormalization flow diagram of the 2CK model (reproduced from [7]). The arrows are pointing towards low temperatures. This diagram shows the renormalization of the exchange coupling J_i of each channel. One can see the finite coupling 2CK fixed point and the 1CK fixed points (at infinity). 'LM' is the local-moment fixed point of an uncoupled impurity ($J_1 = J_2 = 0$)

Note that in the 3CK model, a channel anisotropy $J_1 = J_2 > J_3$ drives the system from the 3CK NFL fixed point to the 2CK NFL fixed point. Therefore, a channel anisotropy does not necessarily generate a crossover towards a FL ground state. This is in contrast with an energy level splitting ΔE. We will experimentally explore both of these type of crossovers (using ΔJ and ΔE separately).

Crossover temperature $\mathbf{T_{co}}$

As shown in Fig. 4.2, a power law $T_{co} \propto |g - g_c|^\alpha$ delimits the quantum critical region from the Fermi liquid phase. In the multi-channel Kondo effect, the non-thermal parameter g can be either the channel asymmetry ΔJ or the effective magnetic field ΔE (or a combination).

Cox and Zawadowski give the general expression of T_{co} for a channel or magnetic field perturbation (see Sect. 5.1.4. in their extensive review [24]). We will only consider their result on the effective magnetic field ΔE, for the N-channel Kondo model (with $N > 1$):

$$T_{co} \propto \Delta E^\beta \text{ with } \beta = \frac{2+N}{N} \tag{4.2}$$

For a channel symmetry perturbation, the crossover temperature is given by

$$T_{co} \propto \Delta J^\eta \text{ with } \eta = \frac{2+N}{2},$$

Note that the two exponents β and η are equal only for $N = 2$.

Crossover from criticality in tunable devices

In 2015, Keller and co-workers have observed the ΔJ driven crossover from quantum criticality in a 2CK experiment [6]. In their 'spin' implementation, a true magnetic field, a channel asymmetry or coherent charge transfers between the reservoirs are relevant perturbations. They have used a lateral gate voltage (named V_{BWT}) to break the symmetry of the channel couplings to their quantum dot. In Fig. 4.5a, traces of the conductance versus voltage bias (V_{SD}) are plotted. The V_{BWT} value used (vertical dashed line in Fig. 4.5b) is slightly away from the QCP. Therefore, $T_{co} > 0$ and the system is experiencing quantum criticality only at high temperatures $T > T_{co}$. The deviation between the low temperature traces and the universal 2CK predicted behavior (solid black line) signal $T \lesssim T_{co}$.

A crossover temperature scale T_{co} can be extracted from the fit of the conductance traces with a 2CK prediction. This temperature scale is shown in Fig. 4.5b with respect to V_{BWT}. In the 2CK model, we know from the previous paragraph that $T_{co} \propto \Delta J^2$. The red lines of this figure are thus fits to piecewise[4] parabolas that delimit the quantum criticality area (above the parabolas).

One may notice that the power law used in Fig. 4.5a to rescale the experimental data is $\propto \sqrt{T}$ rather than the T-linear power law we have observed in our 'charge'

[4] It is argued in [6] that in general, the prefactor of a critical power law as $T_{co} \propto \Delta J^2$ can be different on each side of a quantum critical point. The reason for such an anisotropy is unclear to us.

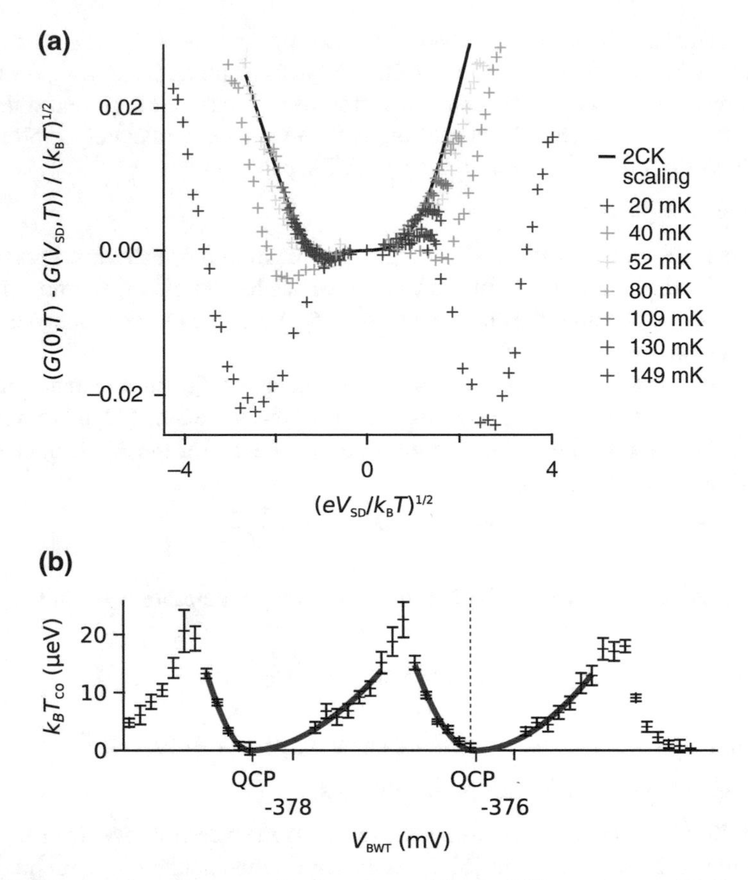

Fig. 4.5 Crossover from quantum criticality observed in the 2CK experiment of Keller and co-workers [6]. **a** The rescaled conductance is plotted versus the rescaled voltage bias. The high temperatures traces follow the universal behavior (black line) whereas at low temperature deviations are observed. **b** A crossover temperature $T_{\rm co}$ can be extracted from a fit of the conductance shown in **a** to a 2CK prediction. This $T_{\rm co}$ is then fitted with half parabolas (that are periodic just as the Coulomb oscillations). The vertical dashed line indicates the voltage gate $V_{\rm BWT}$ used in **a**

2CK measurement. This predicted difference can be intuitively understood from the fact that the conductance G of the 2CK fixed point in the 'charge' implementation is reached at the extremal value $G(J^*_{\rm 2CK}) \triangleq G^*_{\rm 2CK} = e^2/h$. Consequently, from a second-order development $\frac{{\rm d}G}{{\rm d}J}(J = J^*_{\rm 2CK}) = 0$ and one naively expects $G(J(T \ll T_K)) \approx G(J^*_{\rm 2CK} + \alpha\sqrt{T}) \approx G^*_{\rm 2CK} + 1/2\frac{{\rm d}^2G}{{\rm d}J^2}(J^*_{\rm 2CK})(\alpha\sqrt{T})^2$. Thus $|G(T \ll T_K) - G^*_{\rm 2CK}| \propto T$. Note that for $N \geqslant 3$, the same standard power law versus the temperature is expected for the conductance in both the 'spin' and the 'charge' implementations.

4.2.2 Theoretical Predictions for the 'Charge' Kondo Model

Universality in the vicinity of the 2CK fixed point

Considering the 2CK model is very informative since the full quantitative prediction of the 2CK conductance has been computed by Matveev and Furusaki near the 2CK fixed point. They have found the following expression for the conductance along the crossover from a fully developed quantum criticality ($T \ll T_K$) :

$$\tilde{G}_{2\mathrm{CK}}^{T \ll T_K}(T, \Delta\tau, \Delta E) = G_K \left[1 - \frac{T_{\mathrm{co}}(\Delta\tau, \Delta E)}{2\pi T} \psi' \left(1/2 + \frac{T_{\mathrm{co}}(\Delta\tau, \Delta E)}{2\pi T} \right) \right] \tag{4.3}$$

where ψ is the digamma function. Quite generally, in the vicinity of a Kondo quantum critical point (i.e. once $T \ll T_K$), the only energy scale to consider is T_{co} and observables can be expressed as universal functions of T/T_{co} [25–27]. This expression has been derived with another analytical method by Mitchell and co-workers, starting from the opposite limit $\tau \ll 1$, and for both magnetic field and channel asymmetry perturbations [27]. For $N = 2$ channels, the critical exponents associated to the two relevant perturbations are the same: $T_{\mathrm{co}} \propto \Delta\tau^2$ and $T_{\mathrm{co}} \propto \Delta E^2$; and they have the following expression for the crossover temperature [27]:

$$T_{\mathrm{co}} = c_1 T_K \Delta\tau^2 + c_2 \Delta E^2 / T_K \tag{4.4}$$

where c_1 and c_2 are non-universal parameters (T_{co} itself is defined up to an arbitrary prefactor). Furusaki and Matveev have derived the quantitative prediction for $T_{\mathrm{co}}(\Delta\tau, \Delta E)$ from the Hamiltonian parameters τ, ΔE, E_C. Expending their expression to second order in $\Delta\tau \ll 1 - \tau \ll 1$ and $\Delta E \ll E_C$, one finds [28]:

$$T_{\mathrm{co}} \sim \frac{\pi\gamma^2}{4} T_K \Delta\iota^2 + \frac{4}{\pi^3} (\Delta E / k_B)^2 / T_K \tag{4.5}$$

where $k_B T_K \triangleq 2E_C / (\pi^3 \gamma (1 - \tau))$ is defined as in Eq. (3.11). In the next paragraph, we set $\Delta E = 0$ and explore $\Delta\tau$ on the full range $\tau \in [0, 1]$. The crossover temperature and the universal behaviors in T/T_{co} will be discussed afterwards.

Channel asymmetry perturbation $\Delta\tau \neq 0$

Furusaki and Matveev have proposed a renormalization flow diagram for the conductances in the 'charge' 2CK model and the 2CK model with true spin (a particular case of 4CK model) [28]: the channel with the lowest initial bare transmission will end to zero conductance whereas the best initially coupled channel will flow to the $(N/2)$-CK fixed point (see Fig. 4.6).

Note that the lines of the diagrams are hand sketches between the fixed points (they are not exact calculations). Note also that the 1CK fixed point has not been

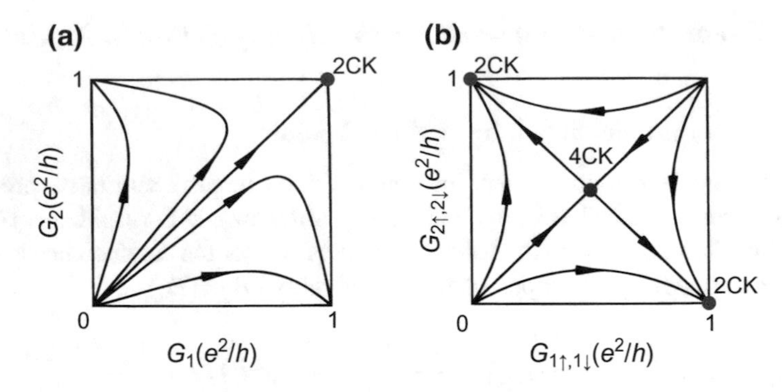

Fig. 4.6 Renormalization flow diagram for the 2- and 4-channel 'charge' Kondo effect (reproduced from [28]). The renormalization flow towards the fixed points when lowering the temperature is indicated by the black arrows. The 4CK model (**b**) corresponds to a 2CK model (**a**) with electrons carrying a true spin. The red and purple dots locate the 2CK, $G^*_{2\text{CK}} = G_K$, and the 4CK, $G^*_{4\text{CK}} = 1/2\,G_K$, fixed points respectively. In the 2CK diagram (**a**), the position of the 1CK fixed point $G^*_{1\text{CK}}$ is not certain (see text)

calculated in [28], and its position at G_K is only postulated by the authors of Fig. 4.6. According to our naive reasoning, the 1CK fixed would rather be at $G^*_{1\text{CK}} = 0$ (see Fig. 3.8).

Theoretical prediction for an effective magnetic field perturbation $\Delta E \neq 0$

Finite ΔE are observed when measuring Coulomb blockade peaks (conductance versus the gate voltage V_g). The shape and the height of the peak depend on the temperature T and the transmissions τ_i of each channel. In Chap. 3 we focused on the height only (the conductance at charge degeneracy $\delta V_g = 0$); here we will consider the full peak, with symmetric transmissions $\tau \triangleq \tau_1 = \tau_2$.

As explained above, the conductance near the Kondo fixed point $T \ll T_K$ is a universal function of T/T_{co} and its expression $G_{2\text{CK}}^{T \ll T_K}$ is given by Eq. (4.3). The full expression of T_{co} (beyond the limit $\Delta E \ll E_C$ given in Eq. (4.5)), has been computed by Furusaki and Matveev for $1 - \tau \ll 1$ [28]:

$$k_B T_{\text{co}} = \frac{8\gamma E_C}{\pi^2}(1-\tau)\sin(\pi \Delta N_g) \tag{4.6}$$

where we use $\Delta N_g \triangleq \delta V_g/\Delta V_g = \Delta E/2E_C$ to simplify. The conductance $G_{2\text{CK}}^{T \ll T_K}$ is plotted in Fig. 4.7a at $\tau = 0.9$ and for different temperatures. Figure 4.7c includes the predicted T-linear deviation from a fully developed quantum criticality (see Eq. (2.9)).

In general, the larger the temperature, the broader the conductance peak. However, in contrast with a standard thermal broadening, here the range of ΔN_g where the conductance is close to $G^*_{2\text{CK}}$ widens with temperature according to a well defined critical exponent. A dashed line is plotted at the crossover conductance $G_{\text{co}} \triangleq 0.5 G_K$

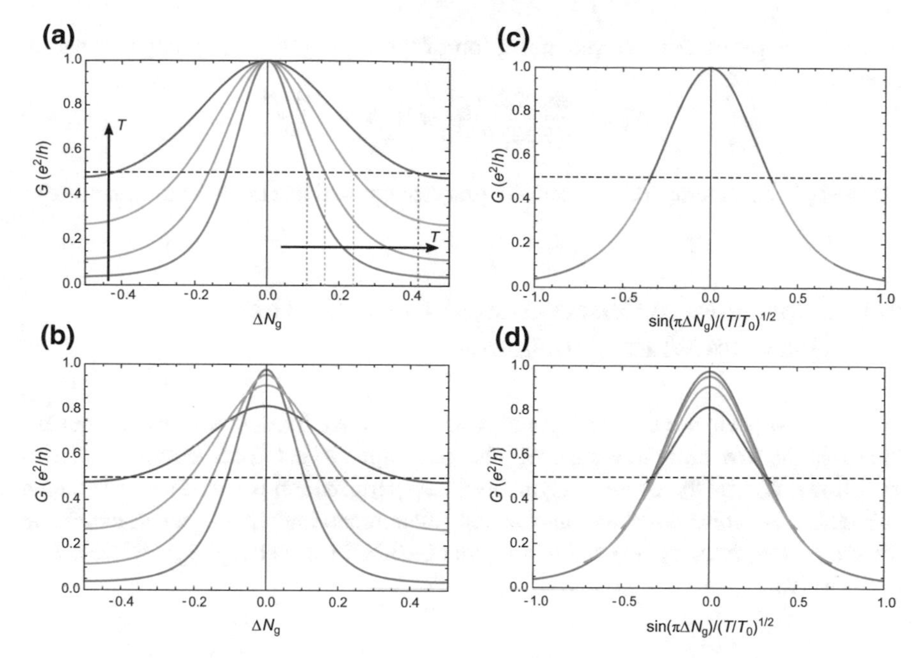

Fig. 4.7 Rescaling of a 2CK Coulomb peak. **a** Plot of the conductance $\widetilde{G}_{2\mathrm{CK}}^{T\ll T_K}$ over one period in ΔN_g for $T = \{5, 10, 20, 40\}$ mK (from bottom to top) with $E_C/k_B = 300$ mK. The black dashed line indicates $G_{\rm co} = 0.5G_K$. The typical width of the quantum critical area is given by the intersection of the conductance G with $G_{\rm co}$, it is shown by the vertical colored short dashed lines. **b** Same as **a** but the T-linear term is taken into account. **c** The conductance traces shown in **a** are rescaled (see the x-axis). **d** Same as **c** but the T-linear term is taken into account

used to characterize this effect. This broadening of the quantum critical area with the temperature should agree with the general expression Eq. (4.2) given by Cox and Zawadowski. For $N = 2$ channels, the crossover should occur on a temperature scale $T_{\rm co} \propto \Delta N_g^2$ for $\Delta N_g \ll 1$.

Let us go further and look for a method to check whether the conductance measured experimentally follows a universal function $G(T/T_{\rm co})$. For $\Delta N_g \ll 1$, $T_{\rm co} \propto \Delta N_g^2$ is a power law; therefore, a method would be to plot $G(T/\Delta N_g^2)$, or equivalently $G(\Delta N_g/T^{1/2})$. As we know the general expression of $T_{\rm co}$, we can use a rescaling that is valid even beyond the limit $\Delta N_g \ll 1$ where $T_{\rm co}$ was a simple power law. We therefore rescale the axis $\Delta N_g \mapsto \sin(\pi N_g)/\sqrt{T/T_0}$, with T_0 a temperature of reference. This rescaling has been applied in Fig. 4.7, where one can observe that the curves of Fig. 4.7a for different temperatures *exactly* collapse on a single rescaled curve in Fig. 4.7c on the full range of ΔN_g.

The same rescaling is tested in Fig. 4.7d on the full expression $G_{2\mathrm{CK}}^{T\ll T_K}$ including the T-linear term describing the deviation from a fully developed quantum criticality. This term vanishes at $\Delta N_g = 1/2$, hence this rescaling works well on the 'tails' of the conductance peaks. A naive generalization of the scaling to the N-channel 'charge' Kondo is:

$$\Delta N_g \mapsto \frac{\sin(\pi \Delta N_g)}{(T/T_0)^{1/\beta}} \text{ with } \beta = \frac{2+N}{N} \tag{4.7}$$

where T_0 is just a temperature scale (in practice we will use the base temperature).

4.3 Experimental Observation of the Multi-channel Quantum Phase Transitions

We will now present our experimental results when we introduce a relevant perturbation in the two- and three-channel 'charge' Kondo effect. This section is divided as follows: (i) first the channel asymmetry $\Delta\tau$ perturbation is discussed, (ii) then a subsection is dedicated to the unexpected phenomena that appear when exploring the channel asymmetry (iii) and finally we consider an effective magnetic field δV_g perturbation.

4.3.1 Development of the Kondo QPT Versus Channel Asymmetry $\Delta\tau$

We have observed a flow towards the predicted fixed points of the two- and three-channel 'charge' Kondo model in Fig. 3.12 of Chap. 3. In that chapter, the transmission had been adjusted to get symmetric channels ($\tau_1 = \tau_3$ and $\tau_2 = 0$ to reach the 2CK fixed point and $\tau_1 = \tau_2 = \tau_3$ for the 3CK). Here we purposely set an asymmetry $\Delta\tau$ between the channel to observe the full renormalization flow of the in-situ conductances of each Kondo channel.

Kondo renormalization flow diagrams

For the 2CK diagram shown in Fig. 4.8, one needs to open three QPCs to extract the individual conductances of each channel, as explained in Appendix C.1. Indeed, with only two non-zero transmissions ($\tau_{1,2} > 0$), one has only access to the serial conductance through the two junctions ($G_1G_2/(G_1 + G_2)$). In this case, we have weakly coupled a third QPC to be able to probe the individual conductances G_1 and G_2. In order to minimize the perturbations on the 2CK physics, we set $\tau_3 \ll \tau_{1,2}$. Note that the axes do not start at zero because the tunnel configurations $\tau_3 \ll \min(\tau_1, \tau_2) \ll 1$ are hard to access experimentally. In this entire diagram, the *in situ* conductance of the probe channel QPC_3 obeys $1/150 < G_3/\min(G_1, G_2) < 1/6$. The error bars are based on the statistics from successive Coulomb peaks.

For the 3CK diagram shown in Fig. 4.9, no probe is required since three channels are involved. In principle, a 3CK flow diagram should be represented in a 3D plot. But this will be too long to acquire, very redundant and hardly readable. Hence we

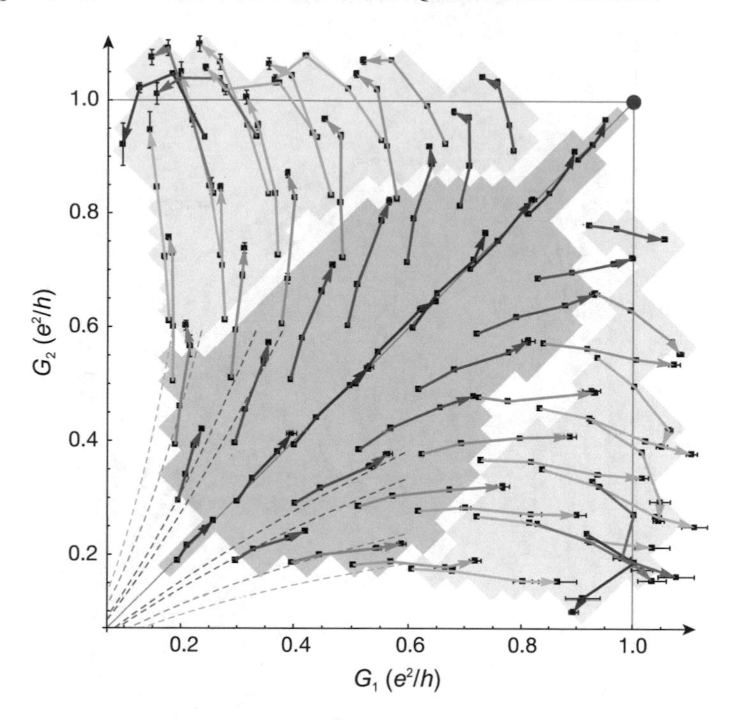

Fig. 4.8 Observation of the 2CK renormalization flow diagram. The in-situ conductances G_1 and G_2 are measured for temperatures $T \approx \{14, 22, 38, 80\}$ mK (using $G_3 \ll G_{1,2}$ as a probe). Each arrow corresponds to a fixed configuration (τ_1, τ_2, τ_3). The arrows point towards the lowest temperature; their color shows the asymmetry ($\Delta\tau \approx 0$ purple, $\Delta\tau \approx 0.57$ red). Errors bars come from statistics on successive Coulomb peaks. The dashed lines represent a *poor man's* scaling. The light gray area delimits the region where the weakest transmitted channel decouples (its conductance decrease). The red disk indicates the location of the 2CK fixed point

decide to set two channels symmetric $\tau_1 = \tau_3$ and plot the data in a 2D graph G_2 versus $G_{1,3}$ (where we have averaged[5] G_1 and G_3). Most of the data in this graph have an uncertainty on both G_2 and $G_{1,3}$ smaller than $0.05G_K$, except the data points connected from high temperatures with a dashed line (whose uncertainty is between $0.05G_K$ and $0.1G_K$, these points are harder to acquire because $G_{1,3}$ is low whereas G_2 is high).

Tunnel case $\tau \ll 1$

In the multi-channel Kondo model, channels weakly coupled to the impurity (when $T \gg T_K$) are renormalized independently. This is shown in the 2CK diagram where

[5]This is possible since they remain approximately equal. The typical difference between G_1 and G_3 can be observed on the diagonal of the diagram, from the similar difference between G_2 and $G_{1,3}$.

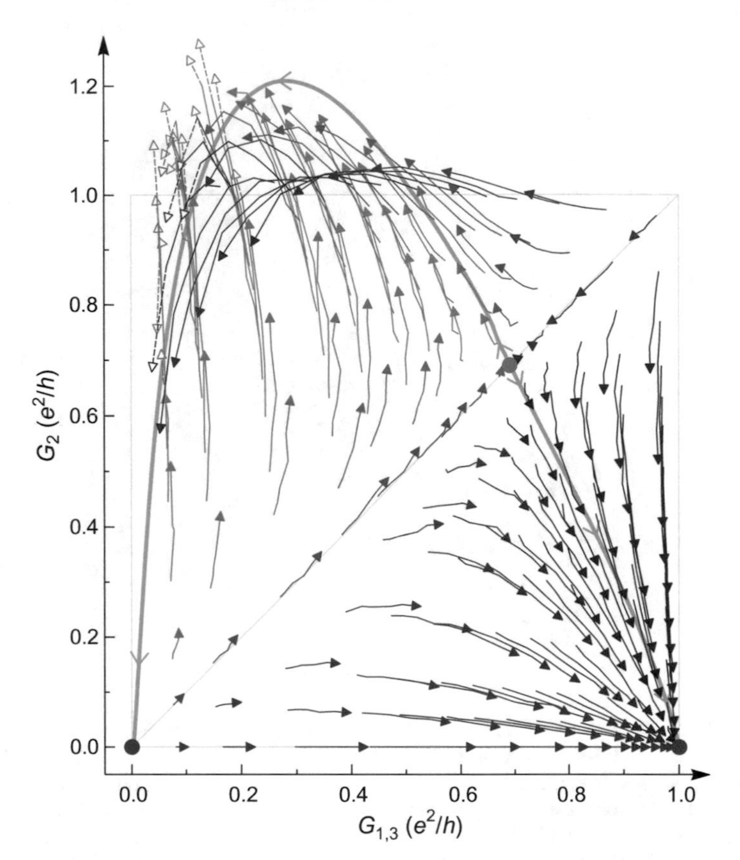

Fig. 4.9 Observation of the 3CK renormalization flow diagram. The averaged conductance of G_1 and G_3 is plotted versus G_2 at temperatures $T \approx \{7.9, 9.5, 12, 18, 29, 40, 55\}$ mK. Here, the color of the arrows maps their orientation. The uncertainty on the open symbols and dashed lines is smaller than $0.1\, G_K$ whereas it is smaller than $0.05\, G_K$ for the solid symbols and lines. The fixed points are indicated with colored dots (1CK in cyan, 2CK in red and 3CK in green). The grey lines are NRG calculations

the dashed lines correspond to the *poor man's scaling* defined[6] by Eq. (3.8). Each line is associated to the arrow it is touching (and which is of the same color). One can see that this approximate scaling works quite well although the transmission are not really tunnel-like.

Symmetric coupling $\Delta\tau = 0$

This case was the topic of Chap. 3 where the flow towards the two- and three-channel Kondo fixed point was studied quantitatively. Here we directly show that symmetric

[6]To be precise, we plot $G_2^{T \gg T_K}(T/T_K(\tau_2))$ versus $G_1^{T \gg T_K}(T/T_K(\tau_1))$, where $G^{T \gg T_K}(T/T_K) = 19, 24\, G_K \ln^{-2}(T/0.0037 T_K)$ is the tunnel regime of the universal curve (see Fig. 3.14a), and where $T_K(\tau_1)$ and $T_K(\tau_2)$ correspond to the $T_K(\tau)$ in the inset of the same figure.

channels remain symmetric under the renormalization flow. Note that the symmetry was fine tuned to symmetric couplings at low temperature $T \approx 18$ mK. The procedure is explained in Appendix C.2.

Development of a quantum phase transition when $\Delta\tau \neq 0$

At zero temperature, the conductances should reach one of the three fixed points depending on the initial setting on the transmissions. Figures 4.8 and 4.9 show the development of a quantum phase transition near the $\Delta\tau = 0$ line as the temperature is reduced. In the 2CK diagram, the dark gray area contains the points which show an increase of both G_1 and G_2, whereas in the light gray area one of the conductances is decreasing.

In the 3CK diagram, the influence of the different fixed point (marked with colored dots) is even more obvious. Moreover it displays two quantum critical points (the 2CK and the 3CK fixed points). One can then observe a remarkable crossover from a NFL fixed point (the 3CK) to another NFL fixed point (the 2CK).

4.3.2 Unanticipated Features of the Flow Diagrams

The renormalization flow diagrams shown in Figs. 4.8 and 4.9 display some previously unanticipated features: Whereas the flow to the two- and three-channel Kondo fixed points appears regular, the flow towards the one-channel Kondo state shows an overshoot of the in situ conductance above the free electron quantum limit and an unanticipated conductance at the 1CK fixed point $G^*_{1CK} = 0$. We point out that these striking features are corroborated by novel NRG calculations.

The 'charge' 1CK fixed point $\mathbf{G^*_{1CK} = 0}$

In the previous chapter, we discussed a naive reasoning to get $G^*_{1\mathrm{CK}} = 0$ from a non-monotonous relation between J and τ (see Fig. 3.8 on page 66). This position of the 1CK fixed point contradicts the diagram sketched by Furusaki and Matveev (see Fig. 4.6a).

The 1CK limit is not visible in the 2CK diagram because we need $G_3 \ll \min(G_1, G_2)$ (in order to consider QPC_3 as non-invading probe). However, the signal-to-noise ratio is good enough in the 3CK diagram in the limit $\tau_{1,3} \ll \tau_2$. We observe that the arrows in this limit (the ones starting from the top of the diagram) are pointing towards zero-conductance at low temperature. Moreover, some arrows in the region $G_{1,3} \approx 0.2$ show a non-monotonous behavior of G_2 (at large G_2).

These two observations are in agreement with the naive picture we propose. They are also supported by NRG calculations [29] as visible on the upper grey line in Fig. 4.9.

Renormalization flow diagram with $\tau_2 = 1$

The limit $\tau_2 = 1$ can be reached by setting the QPC_2 in the middle of the first IQHE plateau. The transmission in this case is perfectly equal to one within our

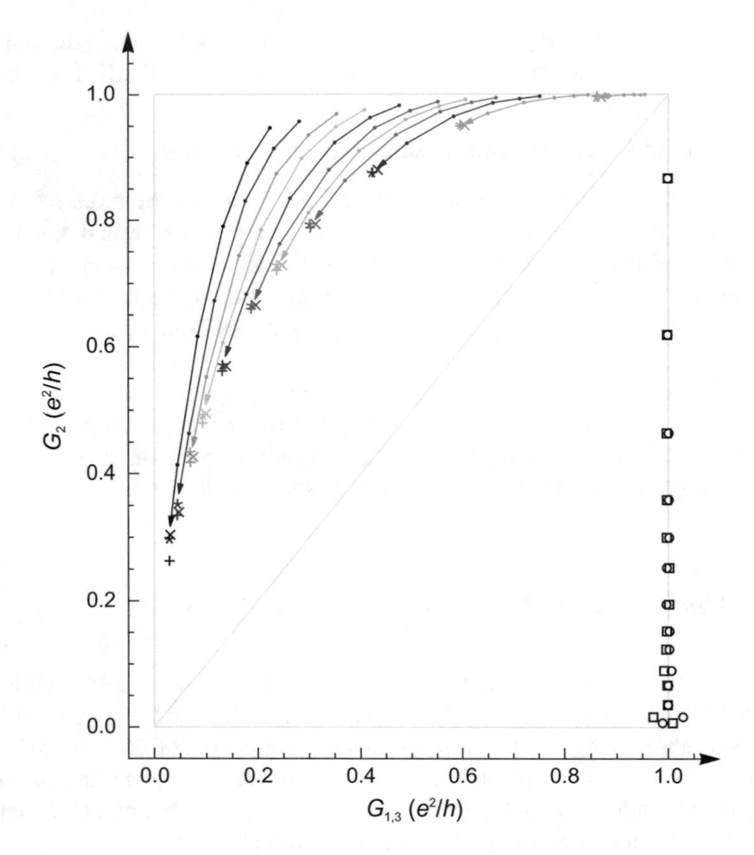

Fig. 4.10 Renormalization flow of the in-situ conductances when some transmissions are set at $\tau = 1$. Each point is an average over δV_g (see text). The arrows show the in-situ conductances G_2 versus $G_{1,3}$ at $\tau_2 = 1$ for $T \approx \{8.9, 9.5, 12, 18, 29, 40\}$ mK. For comparison, the role of QPC$_2$ has been replaced by QPC$_1$ (cross), QPC$_2$ (plus) and QPC$_3$ (star) at $T \approx 8.0$ mK. The squares and the circles are measured at $T \approx 8.0$ mK only, at $\tau_1 = \tau_3 = 1$ for different values of τ_2. The squares correspond to G_1 and the circles to G_3

measurement accuracy (see the discussion on the reflectivity at the 2DEG-metallic island interface in Appendix A.2). In this case, the transmission of the channel is said *ballistic*, and one naively expects a conductance G_2 equal to the quantum of conductance G_K whatever the environment of QPC$_2$ (i.e. whatever τ_1, τ_3 and E_C) and whatever the temperature[7] (below a reasonable limit).

In Fig. 4.10 we present the flow diagram for $\tau_2 = 1$ (with $\tau_1 = \tau_3$). As we have demonstrated in Chap. 2, there is no more charge quantization at $\tau_2 = 1$. Hence the charge degeneracy $\delta V_g = 0$ has no more meaning. What is plotted in this diagram is

[7]For example, see the Fig. 3 of [30], where the conductance of a QPC in series with an ohmic resistor of resistance $R \approx 26\,\mathrm{k}\Omega \approx R_K$ does not drop at low temperature in the limit of ballistic transmission $\tau_\infty \longrightarrow 1$.

the conductance averaged while sweeping V_g (to increase the signal-to-noise ratio). We observe a clear drop of the conductance G_2 when lowering the temperature!

In the same Fig. 4.10, the circle and square symbols correspond to the 'reciprocal' situation where $\tau_{1,3} = 1$ (with $0 < \tau_2 < 1$) at the base temperature only. In this case we observe that both G_1 and G_3 remain close to G_K (at our experimental accuracy).

The drop in the in-situ conductance G_2 is quite surprising since one would naively expect that a quantum conductor at unitary transmission $\tau = 1$ behaves as a classical resistor with a resistance $R = R_K$. In particular, the quantum shot noise is known to vanish at this value of the transmission as for a classical resistance. We see that this picture matches the data when two channels are set to $\tau = 1$ but not with a single channel.

We have tried to substitute the role of QPC_2 by QPC_1 or QPC_3 (see the 'cross', 'plus' and 'star' symbols in Fig. 4.10). We made the measurement only at base temperature and we have observed that the conductance of the channel which is set to $\tau = 1$ does not depend on the QPC. This observation shows that the conductance drop cannot be explained with a residual reflection due to a τ strictly lower that one. Indeed, the residual reflection would not be exactly the same for the three QPCs.

Arrow crossings and universality

In the 3CK renormalization flow diagram, we observe many arrow crossings (in the upper part $\tau_2 > \tau_{1,3}$). These crossings also exist in the 2CK diagram, but are less spectacular, barely above noise. They are in an *apparent* contradiction with the universal behavior expected at low temperature $k_B T \ll E_C$. Indeed, if $(G_2; G_{1,3})$ fully parametrize the universal flow, the trajectories cannot cross.

However, these crossings do not necessarily imply a non-universal behavior if G do not completely characterize the renormalization flow. We expect it to be the case as the relation between J and G is non-monotonous in absence of interaction (see Fig. 3.8). Therefore, a universal flow diagram with the parameter J would contain crossings when plotted versus G.

In the lower part ($\tau_2 < \tau_{1,3}$) of the diagram, the arrows do not cross each other, at the measurement accuracy (in agreement with the monotonous relation between G and J expected along the flow to the 2CK fixed point, which is located at the maximum of the function $G(J)$).

Overstepping the quantum of conductance

The observation of an *in situ* conductance of a single channel larger than the quantum of conductance e^2/h is a striking property of both the 2- and 3-CK flow diagrams. It is especially clear in the 3CK measurement where the signal-to-noise ratio is better. This observation relies on fine calibrations of the offset and the gain of the conductance measurement (see Appendix C.12). This allowed us to plot data with accuracy better than $0.05\,G_K$ (plain arrows) and avoid any doubt about the actual overstepping of G_K.

One should also notice that the overstepping of G_2 above G_K increases as τ_2 is lowered. Note also that there is no overstep in the opposite limit $\tau_2 = 1$ (see Fig. 4.10).

From a general point of view, it is certain that no more than a single quantum Hall edge channel is involved at each QPC since the voltages applied on these split gates completely reflect the other channels. The way we extract the in situ conductances from the classical conductance composition laws is explained in Appendix C.1. Importantly, the *measured* conductance (per channel) through the whole device does not overstep the quantum limit G_K because of the low conductance of QPC_1 and QPC_3.

The NRG calculations of Mitchell corroborate these observations, meaning that this effect is included in the Kondo model. However, to date, we have no simple picture to explain this property.

4.3.3 Crossover from Criticality Versus Detuning from Charge Degeneracy δV_g

We now set $\Delta\tau = 0$ and explore the perturbation due to an effective magnetic field δV_g. In practice, this consists in measuring the conductance per channel G versus the plunger gate V_g over one period $\Delta V_g \approx 0.70\,\text{mV}$.

Quantum criticality in the Kondo renormalized conductance peaks

Let us start with checking that the parameter space of quantum criticality grows with the temperature, according to the predicted power law. A conductance peak in the 2CK regime at $\tau \approx 0.90$ is plotted for different temperatures in Fig. 4.11. This figure has to be compared to Fig. 4.7a and b.

Figure 4.11 contains a double x-axis: one can read both (i) the δV_g applied in practice (bottom) and (ii) the corresponding splitting in energy level when breaking the charge degeneracy $\Delta E \triangleq 2E_C \delta V_g / \Delta V_g$ (top).

This peak broadening reflects the widening range of ΔE for quantum criticality as the temperature is increased. It can be characterized by the intersection of the conductance with G_{co} (black horizontal line): this crossing occurs by ‘definition’ at $T = T_{\text{co}}(\Delta E)$. In the inset of Fig. 4.11, we have plotted this crossover temperature T_{co} (up to 29 mK) versus ΔE and compared it to the power law $T_{\text{co}} \propto \Delta E^2$ expected for the 2CK effect.

Observation of a power law on the crossover temperature T_{co}

In Fig. 4.12 we have generalized the procedure shown in Fig. 4.11 to other transmissions and also to the 3CK measurement. We observe deviations from the predicted power laws at high temperatures. We know from Chap. 3 that non-universal behaviors occur at temperatures higher than $T \approx 12\,\text{mK}$ (see Fig. 3.14). Therefore, we have distinguished the three lowest temperatures in this figure (solid versus open symbols). The inset shows an extraction of the critical exponent β based on a fit of the data at $T \leq 12\,\text{mK}$ to a power law ΔE^{β}. When we average over all the transmissions, we find $\beta = 1.94 \pm 0.04$ in the 2CK regime (the red line is at the value $\beta = 2$ predicted for 2CK) and $\beta = 1.70 \pm 0.07$ in the 3CK regime (the green line is at $\beta = 5/3 \approx 1.667$,

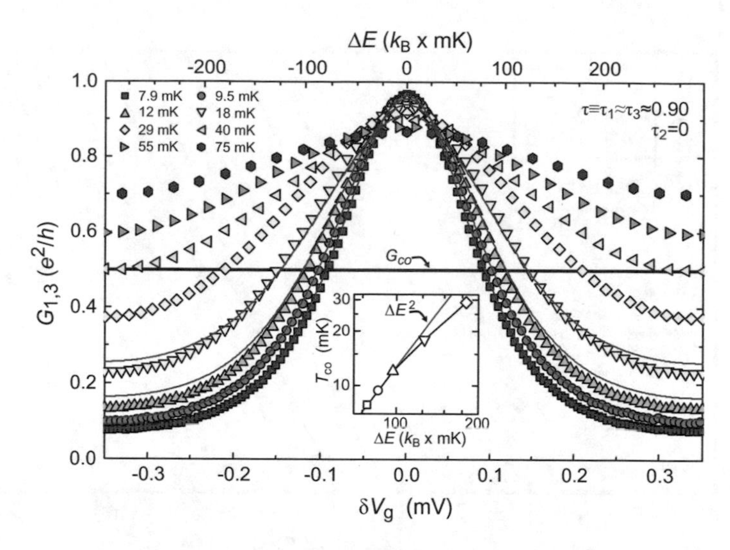

Fig. 4.11 Quantum criticality in a 2CK conductance peak. Coulomb peaks are displayed on a full period $\Delta V_g = 0.70\,\text{mV}$ for different temperatures (each symbol corresponds here to a temperature, see the legend) in the 2CK regime. The red lines are the prediction Eq. 4.3 plotted for the independently measured parameters. The horizontal black line is placed at $G_{\text{co}} \triangleq 0.5\, G_K$. The intersection of $G_{1,3}$ with this line gives the $T_{\text{co}}(\Delta E)$ shown in the inset as symbols. The red straight line corresponds to the 2CK power law $T_{\text{co}} \propto \Delta E^2$, it is adjusted to match the lowest temperature (square) symbol

as predicted for 3CK). Although these two power laws are relatively close, we are clearly able to discriminate between the 2CK and 3CK predictions.

Universality of the conductance expressed in $\mathbf{T/T_{co}}$

We can go further and verify that once the temperature has been lowered enough to reach the non-Fermi liquid regime ($T \ll T_K$), the only energy scale to consider is the crossover temperature T_{co} (provided $k_B T \ll E_C$).

As it has been explained in Sect. 4.2.2, we have a method to check whether the conductance follows a universal function of T/T_{co} on a full range of ΔE. The *ad hoc* function to use is $\Delta E \mapsto \sin(\pi \Delta E/(2E_C))/(T/T_{\text{base}})^{1/\beta}$ where $T_{\text{base}} = 7.9\,\text{mK}$ is the base temperature of the experiment (see Eq. (4.7)). Using this rescaling, we observe in Fig. 4.13a and b (see Fig. 4.13c and d for an updated version using an equivalent rescaling in T_{co}/T) that the conductance peaks at all temperatures collapse on a single curve. For the two-channel case, this curve matches well the theoretical prediction of Furusaki and Matveev given by Eq. (4.3) [28] (see also Fig. 4.7). Note that this prediction is analytical, it was plotted at zero temperature without any fit parameter. For the three-channel case, we observe a good agreement between the T_{co}/T-rescaled measurements and the recent NRG calculations of Mitchell where we used the prefactor in T_{co} as an adjustable parameter.

The updated figures (Fig. 4.13c and d) also display a dashed-dotted line. Indeed, the conductance Coulomb peak G being symmetric versus δV_g, its series expansion

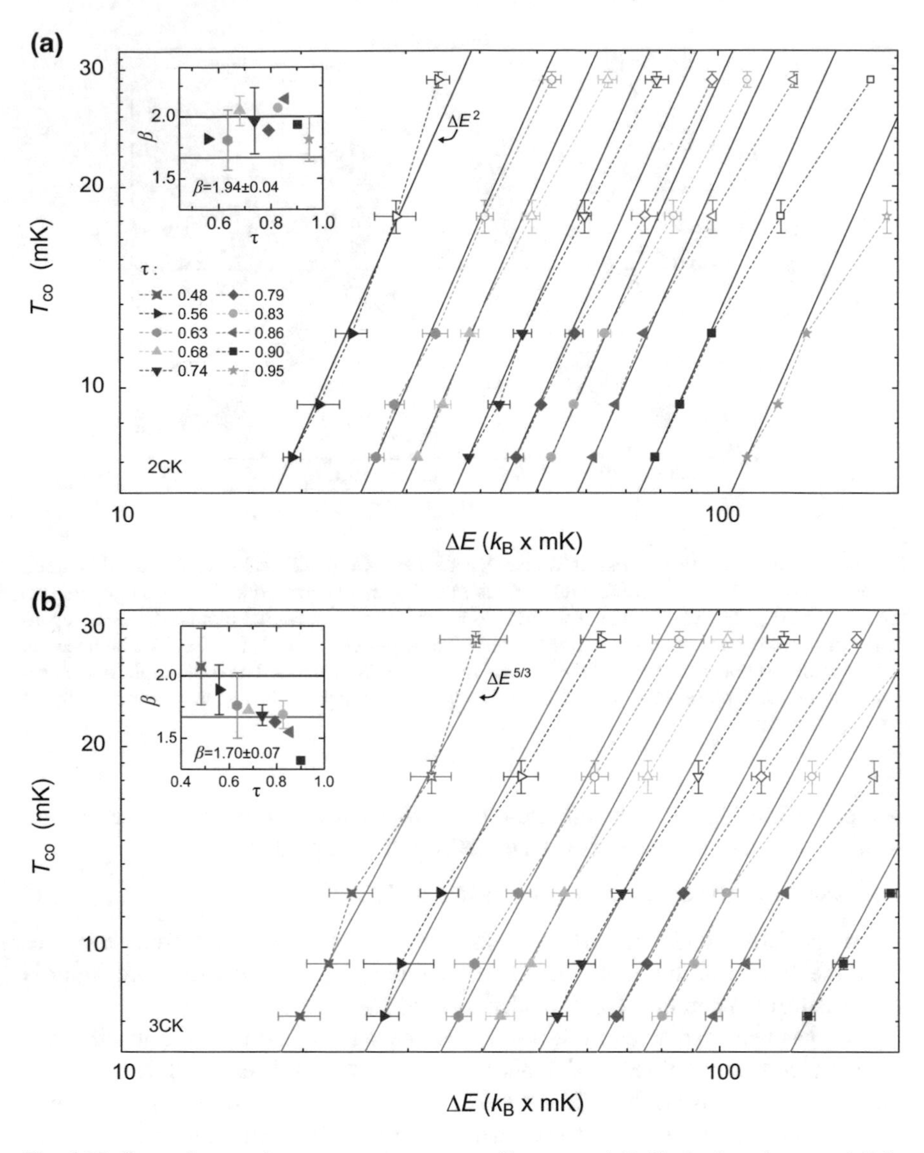

Fig. 4.12 Power law on the crossover temperature T_{co} versus ΔE. T_{co} is plotted versus ΔE for different transmissions (see legend in **a**) in the 2CK (**a**) and 3CK (**b**) regime. The predicted power law (red for 2CK, green for 3CK) is adjusted to match the lowest temperature $T_{base} = 7.9$ mK. In the inset of each graph, we show a fit of the critical exponent β for each transmission τ for the three lowest temperatures (shown as solid symbols in the main graphs). The predicted exponents ($\beta = 2$ in red for the 2CK and $\beta = 5/3$ in green for the 3CK) are plotted for comparison with the mean value obtained over all the transmissions τ

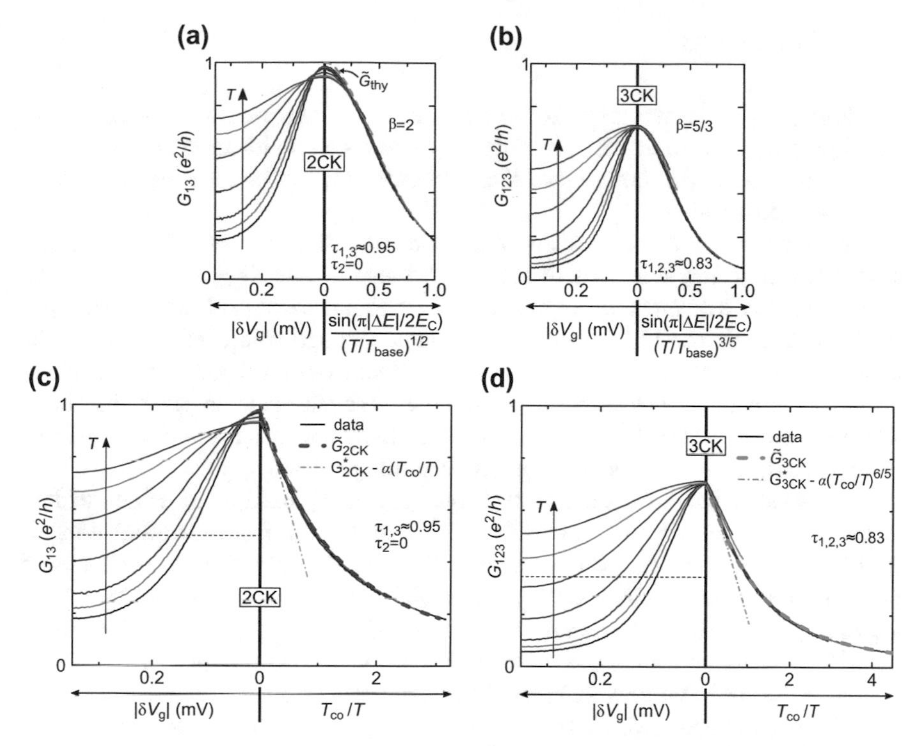

Fig. 4.13 Rescaling of a conductance peak. (old version) a, b, A conductance peak is shown at $T \approx \{7.9, 9.5, 12, 18, 29, 40, 55\}$ mK for a selected transmission τ close to the 2CK (in **a**) and the 3CK (in **b**) fixed points. The solid lines are experimental data. For each graph **a** and **b**, two scales are used: the raw plunger gate voltage δV_g (left) and a rescaled axis $\sin(\pi\Delta E/(2E_C))/(T/T_{\text{base}})^{1/\beta}$ (right). The conductance adopts a universal behavior in the rescaled axis (except for the highest two temperatures). The grey dashed line in **a** shows the zero temperature analytical prediction for $\tau = 0.95$. **(updated version) c, d,** Here, the rescaled axis is T_{co}/T, with $T_{\text{co}} \propto \delta V_g^{\beta}$. The $\tilde{G}_{2\text{CK}}$ red dashed line is **c** the same as the grey in **a**. In **d**, $\tilde{G}_{3\text{CK}}$ is an NRG prediction (where the proportional constant in T_{co} is a fit parameter). Dashed-dotted grey lines are $(T_{\text{co}}/T)^{2/\beta}$ power-laws for $T_{\text{co}}/T \ll 1$

contains only even powers. To the lowest order, $\Delta G(\delta V_g \ll \Delta V_g) \triangleq |G_{1(,2),3} - G^*_{(2)3\text{CK}}| \propto \delta V_g^2$. Using $T_{\text{co}} \propto \delta V_g^{\beta}$, we deduce that $\Delta G(T_{\text{co}}/T \ll 1) \propto (T_{\text{co}}/T)^{2/\beta}$, this is the straight line power-law we plotted.

In order to observe the crossover from quantum criticality, the displayed conductance peak are close to the 2CK and 3CK fixed points. Therefore, their height does not depend much on the temperature (except for the red and the orange traces that deviate from the universal regime defined by $T \ll E_C$). With the scalings showed in Fig. 4.13, we verified that in the vicinity of the two- and three-channel Kondo quantum critical points, the conductance follows a universal function of T_{co}/T on the full range of ΔE.

4.4 Conclusion

In this chapter, we have studied the finite temperature signatures of the second-order quantum phase transition that occurs at the overscreened multi-channel Kondo fixed points. We have used in turn a channel asymmetry and an effective magnetic field as a relevant perturbation.

On the one hand, breaking the channel symmetry allowed us to observe the full renormalization flow diagrams of the conductance. These diagrams illustrate the development of a quantum phase transition as the temperature is lowered. The convergence towards the 1CK fixed point displays a remarkable overstepping of the quantum of conductance e^2/h by the in situ conductance of a single electronic channel (in agreement with novel NRG calculations). We have also observed that the in situ conductance of an electronic channel accurately set at transmission $\tau = 1$ can be reduced towards zero under the renormalization process.

On the other hand, considering the voltage gate V_g used to measure conductance peaks as an effective magnetic field, a crossover from quantum criticality ($\delta V_g = 0$ at $T \ll T_K$) can be studied. The broadening of these conductance peaks is in agreement with a temperature scale that follows a universal prediction $T_{\text{co}} \propto \delta V_g^{(2+N)/N}$ valid whatever the nature of the degrees of freedom used to implement the N-channel Kondo model. More generally, beyond the limit of small δV_g, we have observed that both 2CK and 3CK Coulomb peaks for different temperatures T collapse onto a single curve when plotted versus a rescaled expression of $\delta V_g \mapsto \sin(\pi \delta V_g/\Delta V_g)/(T/T_{\text{base}})^{1/\beta(N)}$. These T/T_{co}-rescaled data are in agreement with analytical and NRG predictions on the full range of δV_g using either no or a single fit parameter.

References

1. S. Sachdev, B. Keimer, Quantum criticality (2011). arXiv preprint arXiv:1102.4628
2. P. Gegenwart, Q. Si, F. Steglich, Quantum criticality in heavy-fermion metals. Nat. Phys. **4**(3), 186–197 (2008)
3. B. Keimer, S.A. Kivelson, M.R. Norman, S. Uchida, J. Zaanen, From quantum matter to high-temperature superconductivity in copper oxides. Nature **518**(7538), 179–186 (2015)
4. H.T. Mebrahtu, I.V. Borzenets, D.E. Liu, H. Zheng, Y.V. Bomze, A.I. Smirnov, H.U. Baranger, G. Finkelstein, Quantum phase transition in a resonant level coupled to interacting leads. Nature **488**(7409), 61–64 (2012)
5. H.T. Mebrahtu, I.V. Borzenets, H. Zheng, Y.V. Bomze, A.I. Smirnov, S. Florens, H.U. Baranger, G. Finkelstein, Observation of Majorana quantum critical behaviour in a resonant level coupled to a dissipative environment. Nat. Phys. **9**(11), 732–737 (2013)
6. A.J. Keller, L. Peeters, C.P. Moca, I. Weymann, D. Mahalu, V. Umansky, G. Zaránd, D. Goldhaber-Gordon, Universal Fermi liquid crossover and quantum criticality in a mesoscopic system. Nature **526**, 237–240 (2015)
7. M. Vojta, Impurity quantum phase transitions. Phil. Mag. **86**(13–14), 1807–1846 (2006)
8. M. Vojta, Quantum phase transitions. Rep. Prog. Phys. **66**(12), 2069 (2003)
9. J. Cardy, *Scaling and Renormalization in Statistical Physics*, vol. 5 (Cambridge University Press, 1996)

10. K.G. Wilson, The renormalization group: critical phenomena and the Kondo problem. Rev. Mod. Phys. **47**(4), 773–840 (1975)
11. J. Zinn-Justin, Critical phenomena: field theoretical approach. Scholarpedia **5**(5), 8346 (2010)
12. S. Martin, A.T. Fiory, R.M. Fleming, L.F. Schneemeyer, J.V. Waszczak, Normal-state transport properties of Bi_{2+x} Sr_{2-y} $CuO_{6+\delta}$ crystals. Phys. Rev. B **41**(1), 846 (1990)
13. J. Custers, P. Gegenwart, H. Wilhelm, K. Neumaier, Y. Tokiwa, O. Trovarelli, C. Geibel, F. Steglich, C. Pépin, P. Coleman, The break-up of heavy electrons at a quantum critical point. Nature **424**(6948), 524–527 (2003)
14. P. Coleman, A.J. Schofield, Quantum criticality. Nature **433**(7023), 226–229 (2005)
15. H.V. Löhneysen, T. Pietrus, G. Portisch, H.G. Schlager, A. Schröder, M. Sieck, T. Trappmann, Non-Fermi-liquid behavior in a heavy-fermion alloy at a magnetic instability. Phys. Rev. Lett. **72**(20), 3262 (1994)
16. I. Bloch, J. Dalibard, S. Nascimbène, Quantum simulations with ultracold quantum gases. Nat. Phys. **8**(4), 267–276 (2012)
17. A.A. Houck, H.E. Türeci, J. Koch, On-chip quantum simulation with superconducting circuits. Nat. Phys. **8**(4), 292–299 (2012)
18. A. Georges, G. Kotliar, W. Krauth, M.J. Rozenberg, Dynamical mean-field theory of strongly correlated fermion systems and the limit of infinite dimensions. Rev. Mod. Phys. **68**(1), 13 (1996)
19. N. Andrei, C. Destri, Solution of the multichannel Kondo problem. Phys. Rev. Lett. **52**(5), 364 (1984)
20. A.M. Tsvelick, P.B. Wiegmann, Solution of the n-channel Kondo problem (scaling and integrability). Zeitschrift für Phys. B Condens. Matter **54**(3), 201–206 (1984)
21. W.W. Ludwig, Critical theory of overscreened Kondo fixed points. Nucl. Phys. B **360**(2), 641–696 (1991)
22. I. Affleck, A.W. Ludwig, Exact conformal-field-theory results on the multichannel Kondo effect: single-fermion green's function, self-energy, and resistivity. Phys. Rev. B **48**(10), 7297 (1993)
23. R. Bulla, T.A. Costi, T. Pruschke, Numerical renormalization group method for quantum impurity systems. Rev. Mod. Phys. **80**(2), 395 (2008)
24. D.L. Cox, A. Zawadowski, Exotic Kondo effects in metals: magnetic ions in a crystalline electric field and tunnelling centres. Adv. Phys. **47**(5), 599–942 (1998)
25. M. Pustilnik, L. Borda, L.I. Glazman, J. von Delft, Quantum phase transition in a two-channel-Kondo quantum dot device. Phys. Rev. B **69**(11), 115316 (2004)
26. A.K. Mitchell, E. Sela, Universal low-temperature crossover in two-channel Kondo models. Phys. Rev. B **85**(23), 235127 (2012)
27. A.K. Mitchell, L.A. Landau, L. Fritz, E. Sela, Universality and scaling in a charge two-channel Kondo device. Phys. Rev. Lett. **116**(15) (2016)
28. A. Furusaki, K.A. Matveev, Theory of strong inelastic cotunneling. Phys. Rev. B **52**(23), 16676–16695 (1995)
29. Z. Iftikhar, A. Anthore, A.K. Mitchell, F.D. Parmentier, U. Gennser, A. Ouerghi, A. Cavanna, C. Mora, P. Simon, F. Pierre, Tunable quantum criticality and super-ballistic transport in a 'charge' kondo circuit, 08 (2017)
30. F.D. Parmentier, A. Anthore, S. Jezouin, H. Le Sueur, U. Gennser, A. Cavanna, D. Mailly, F. Pierre, Strong back-action of a linear circuit on a single electronic quantum channel. Nat. Phys. **7**(12), 935–938 (2011)

Appendix A
Sample Description and Characterization

Figure A.1 shows a colored micrograph of the sample with the name of each element. The elements outside the image are of macroscopic size. There are two types of connections to the outside world: (i) the white circles (ii) the black circle. The black circle is a usual connection: a conductive wire touches the element on the surface. The white circles are ohmic contacts: an alloy of Au–Ni–Ge is diffused to connect the underlying 2DEG by a thermal annealing. The white circles are thus used for injecting or measuring current while the black circles are used to shape the circuit.

The sample is subjected to a strong magnetic field giving rise to IQHE. In this regime, the current is flowing along chiral edge states. In Fig. A.1, only the outermost channel is shown, and not everywhere in order to lighten the figure. Due to the chiral edge states of the quantum Hall effect, the injection and the measurement points are not shot-circuited and one can for instance measure the signal reflected back from a QPC.

This chapter is divided into a section about the gates on the surface (QPCs and SWs gates) and a section dealing with the central micron-sized ohmic contact. The first section explains how we can implement the different circuits studied in this thesis and it shows our mastery of the quantum conductor embedded in these circuits. The second section shows the good quality of the central ohmic contact which ensures the circuit we implement in practice to be a 'perfect quantum simulation' [1] of the explored phenomena.

A.1 Surface Gates

The principle of field effects to change the conductance of the constriction of the 2DEG below surface gates has been explained in Fig. 1.2 for a QPC. This principle works as well for the switch gates (SWs) used for characterization.

Z. Iftikhar, *Charge Quantization and Kondo Quantum Criticality in Few-Channel Mesoscopic Circuits*, Springer Theses,
https://doi.org/10.1007/978-3-319-94685-6

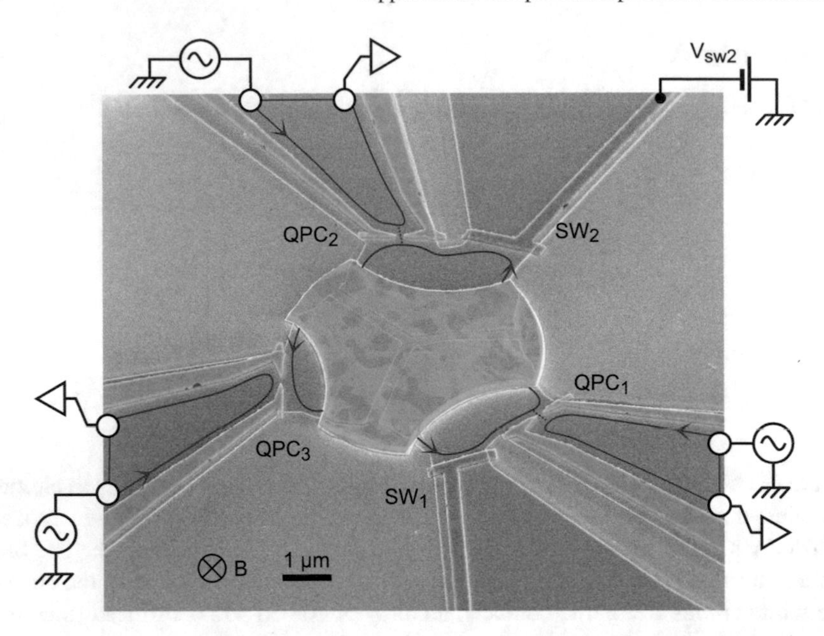

Fig. A.1 Colored micrograph of the sample with measurement schematics. This figures shows the central ohmic contact (in purple) that redistributes the current injected by the a.c. voltage sources (out of the image) into the chiral edge states of the IQHE (red lines) through larger ohmic contacts (white circles) not shown in the picture. The low frequency signal is measured using Lock-in amplifiers (triangles). This sample includes three QPCs (in cyan) and two switch gates (in orange) used for characterization. The value of the transmission of the QPC and the state of the switch is controlled by the voltage sources that connect (black circles) theses gates placed on the surface

A.1.1 Switch Gates and Environment

The switch gates are not used at intermediate transmission because the region right under the gate is generally not clean (one can see the sharp resonances in the gray shaded areas of Fig. A.2). Their geometry is optimized to change the transmission (integer) values quickly, without applying large voltages V_{sw}.

The main purpose of these switch gates is the characterization of the intrinsic conductance of the QPCs.[1] The switches have an injection port, but it is not used to inject current, the edge states lines comes from the ground. When the switch let the current pass, it short-circuits its QPC to the ground. The equivalent circuit is given in Fig. A.3a.

Being in the IQHE regime is not mandatory for the switch gates to work. Indeed, the environment of the QPC will be short-circuited once its lateral switch gate is fully transmitting all the electronic channels. In the IQHE regime, one will measure

[1]There is no switch gate for the third QPC.

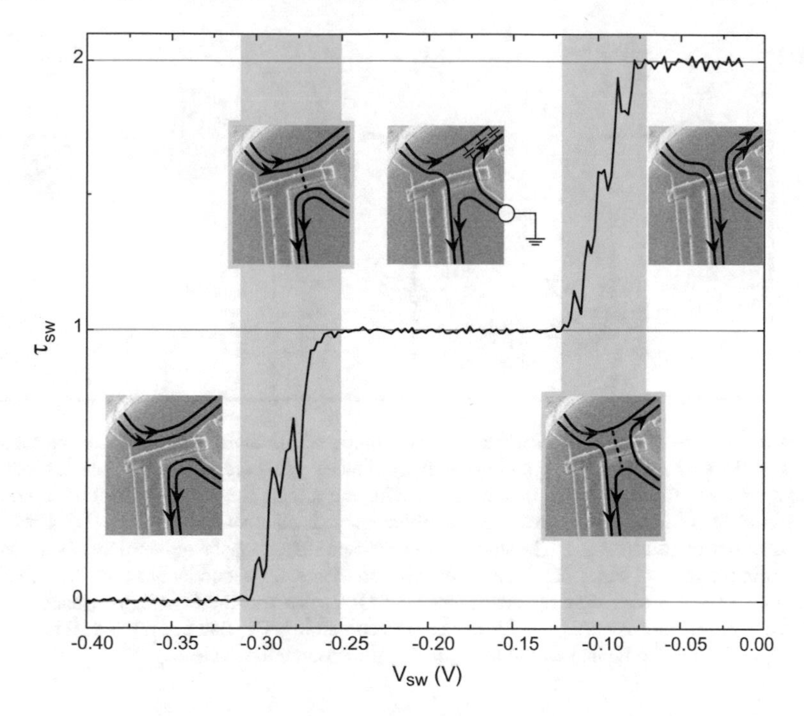

Fig. A.2 A switch gate at work. This figure shows the successive opening of electronic channels by a switch when we sweep the voltage V_{sw}. The insets show the back-scattering of the channels in (the gray shaded areas) and the ballistically transmitted channels (areas without background). The central inset shows two additional information: the switch 'injects' grounded channels and there is a (negligible) capacitive coupling between the edge states

the intrinsic transmission[2] $\tau \leqslant n_{\text{channels}}$ of a QPC (where n_{channels} is the maximum number of channels opened by the QPC for a given characterization) if its lateral switch let pass at least $n_{\text{channels}} \leqslant \nu$.

There might be an interaction (sketched with distributed capacitors in the central inset of Fig. A.2) between the edge states that put some signal in the strict case $n_{\text{channels}} < \nu$, but this is completely negligible on the small distance between a switch and its QPC. In practice all our experiments are done with $\tau \leqslant 1$, and we open at least the switches on $n_{\text{channels}} = 2$.

A.1.2 Characterization and Measurement Circuitry

We have seen that we are able to change the environment of a QPC by changing the state of the switches. Changing the voltage on the switches gate also have an

[2]As explained when discussing Fig. 1.3, in the IQHE regime, we can define transmissions larger than $\tau = 1$ when considering the transmission of more than a single edge channel.

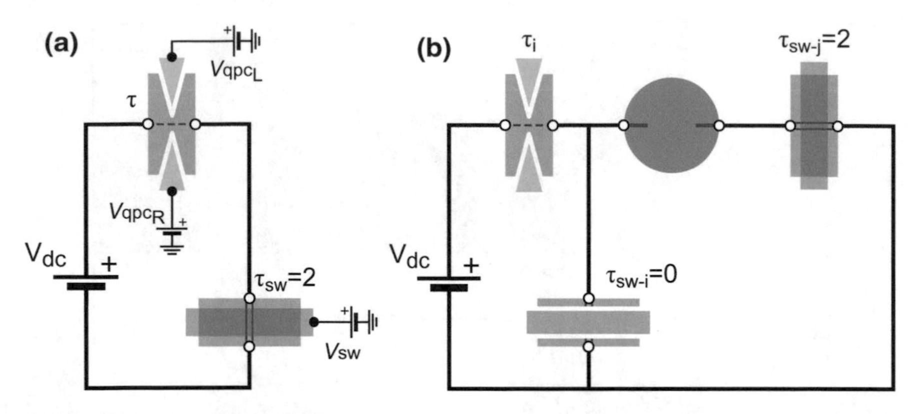

Fig. A.3 Circuits used for characterization. The shape of the 2DEG (in gray) can be modified thanks to the surface gates (in cyan and orange). Theses gates are connected (black circles) to voltage sources. The connection of the 2DEG with the circuit is achieved with ohmic contacts (white circles). **a** Circuit for the measurement of the intrinsic conductance $G_\infty = \tau G_K$. The switch let pass two channels and $\tau \leqslant 1$. The switch is in its closed state, it can be replaced by black classical wire (as in the inset of Fig. 1.3a). **b** In series with an element, the conductance of the QPC_i will be modified $G_i \neq \tau_i G_K$. Here the environment of QPC_i is a resistor $R = R_K/2$ (made with the switch SW_j of transmission $\tau_{sw_j} = 2$). The branch containing SW_i can be erased as it is placed in its open state. This is a typical circuit for Dynamical Coulomb blockade

electrostatic influence on the size of the constrictions of the QPCs. This point will be discussed in the next subsection. Here we discuss different configurations of the switches and the QPCs.

The switch gates are used to implement two characterization circuits. We can use them to short-circuit the environment (Fig. A.3a) or to make a well-controlled resistive environment (Fig. A.3b). The first circuit measures the intrinsic transmission τ of a QPC while the second one probes the conductance in presence of DCB.

The circuit actually studied to get the main results reported in this thesis is given in Fig. A.4. The switches were used for characterization only, they are set to $\tau_{sw_1} = \tau_{sw_2} = 0$ during the experiments. However we use V_{sw_2} as a plunger gate. This gate is swept on a very small range (generally in $[-0.395\,\text{V}, -0.405\,\text{V}]$), and with a good margin to keep $\tau_{sw_2} = 0$ (see Fig. A.2).

A.1.3 Quantum Point Contacts Transmission and Characterization

Let us discuss how we implement a QPC with a well characterized transmission τ. First, this transmission has an energy dependence and its variation $\Delta\tau$ on the range of explored temperatures has to be minimized to guaranty that we were dealing with the same object at the different temperatures. Second, we will play with several QPCs; changing the voltages on the gate of a QPC influences the size of the constriction

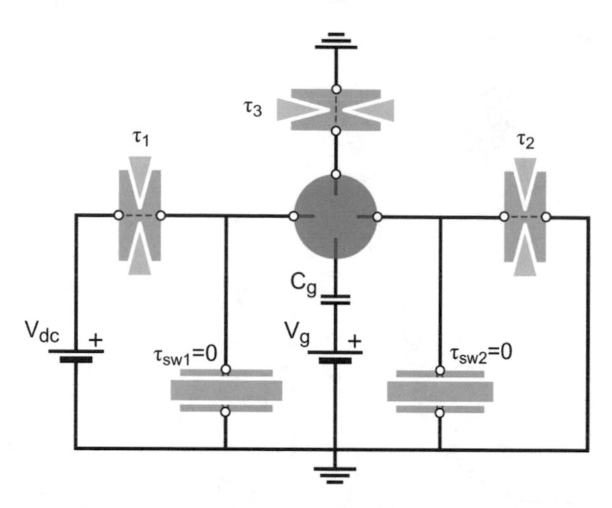

Fig. A.4 Circuit that produced the results reported in this thesis. The branches containing the switches can be erased as both are in opened position. In the charge quantization experiment $\tau_3 = 0$, its branch can be suppressed too. We always have $\tau_1, \tau_2, \tau_3 \leqslant 1$. When $\tau_1, \tau_2, \tau_3 \ll 1$, the charge on the island (in purple) is quantized because it is almost isolated from the circuit. This charge state can be tuned using the voltage V_g applied on a surface gate (e.g. V_{sw_2}) of an element at $\tau = 0$ (not to modify its transmission)

of the other QPCs and therefore their transmission: these crosstalks should be fully characterized. Third, we use the switches to measure the intrinsic transmissions, but we switch their state to perform the real experiment, the crosstalk between a QPC and its lateral switch gate should also be characterized. These two corrections will be discussed in the next subsection.

Quantum point contacts (QPCs) give us access to the full range of transmission $\tau \in [0, 1]$. For the purpose of our experiments, we will pick up a dozen of values in this range. We can check the energy dependence of these selected values of transmission by applying a voltage bias. Figure A.5 shows the transmission of a QPC and the values picked up for the two-channel Kondo experiment (published [2], but not shown in this thesis). The inset shows the energy dependence $\Delta\tau/\tau$ for three of these values versus the voltage bias (the highest temperature in this experiment was $T_{\max} \approx 151\,\text{mK} \approx 13\,\mu\text{V}/k_B$ and the voltage range is $[-50, +50]\,\mu\text{V}$).

Let us verify that our QPCs are generic and interchangeable. A single-channel quantum conductor (as our QPCs in the IQHE regime) is fully characterized by its transmission τ. In Fig. A.6, we have set the three QPCs in turn in the same DCB configuration ($T = 18\,\text{mK}$ and a resistive environment of $R_K/2$ as shown in Fig. A.6a). We have tuned each QPC to a given conductance at zero voltage bias, and we observe in Fig. A.6b that all the DCB curves develop equally (up to some energy dependence) for all the QPCs.

The value of the transmission we have used in Chaps. 3 and 4 are obtained from the conductance in this DCB configuration, at large voltage bias (in the gray shaded areas of Fig. A.6b). The conductance is not the same for positive and negative large

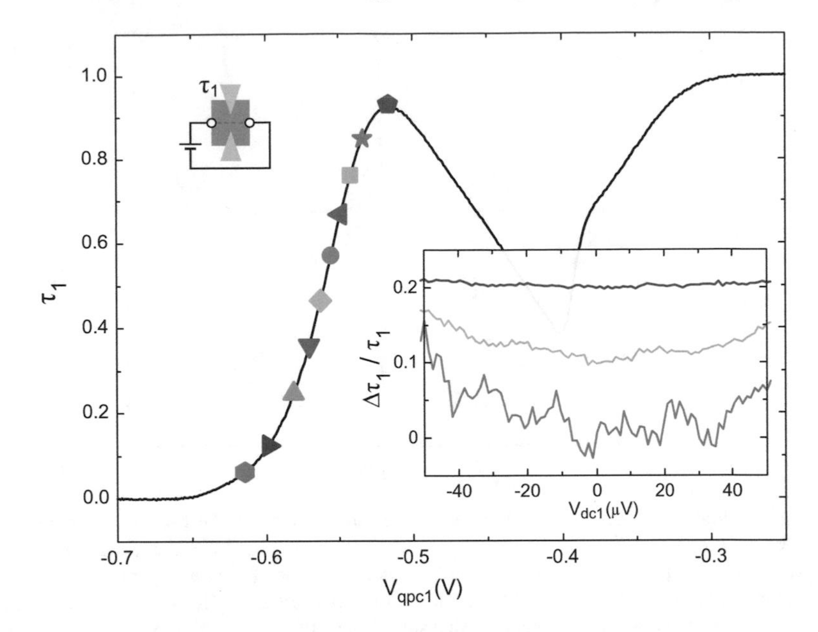

Fig. A.5 Energy dependence of the intrinsic transmission τ. The transmission τ_1 of QPC$_1$ is measured as a function of the voltage V_{qpc_1}. The step from $\tau_1 = 0$ to 1 is marked by a broad resonance around $V_{qpc_1} = -0.4$ V, but we pick up our points (symbols) away from it. The energy dependence is shown in the inset for three selected transmissions, where the difference relatively to $V_{dc} = 0$ is plotted (with an offset for clarity)

bias because of the energy dependence of the QPCs. This limits the accuracy for the determination of a single transmission. The $\Delta\tau$ associated to each transmission τ in the table of Fig. A.6c takes into account both the positive/negative bias and QPC$_{1,2,3}$ scatter (six values for each τ), to get an estimation of the accuracy on τ.

A.1.4 Capacitive Crosstalk Corrections

Changing the voltage on a surface gate has an electrostatic capacitive influence on all the constrictions (QPCs) of the sample. This electrostatic effect depends basically on the distance between the two objects we consider. It can be characterized and corrected by applying a correction to each QPCs gates to compensate the change on a given surface gate.

These corrections are relatively small but they can have a certain impact on the sensitive measurement reported in this thesis. We first discuss the case of distant gates and then treat the crosstalk of a switch gate on its lateral QPC.

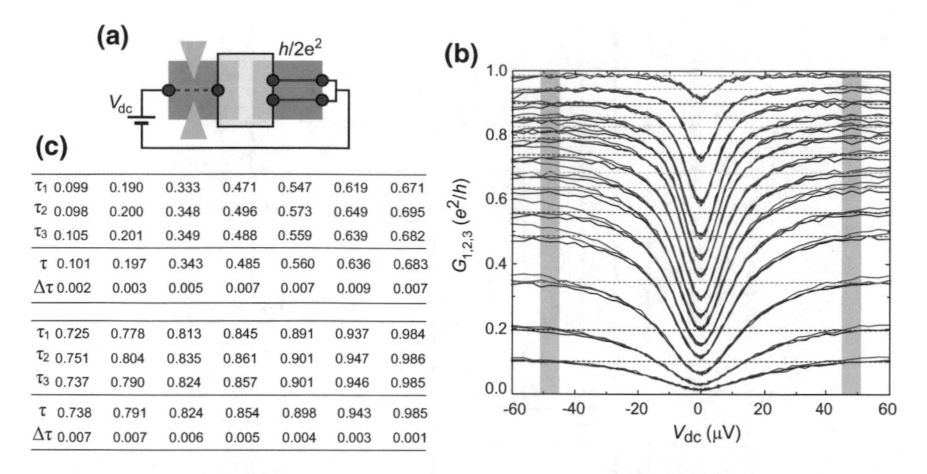

τ_1	0.099	0.190	0.333	0.471	0.547	0.619	0.671
τ_2	0.098	0.200	0.348	0.496	0.573	0.649	0.695
τ_3	0.105	0.201	0.349	0.488	0.559	0.639	0.682
τ	0.101	0.197	0.343	0.485	0.560	0.636	0.683
$\Delta\tau$	0.002	0.003	0.005	0.007	0.007	0.009	0.007

τ_1	0.725	0.778	0.813	0.845	0.891	0.937	0.984
τ_2	0.751	0.804	0.835	0.861	0.901	0.947	0.986
τ_3	0.737	0.790	0.824	0.857	0.901	0.946	0.985
τ	0.738	0.791	0.824	0.854	0.898	0.943	0.985
$\Delta\tau$	0.007	0.007	0.006	0.005	0.004	0.003	0.001

Fig. A.6 Equivalence of QPC in presence of DCB and accuracy on the transmission τ. **a** Schematic circuit used to characterize the transmission probabilities of the QPCs. A lateral characterization gate is used to implement a serial resistance of $R = R_K/2$. **b** The conductance $G^{\text{DCB}}_{1,2,3}$ of QPC$_1$, QPC$_2$ and QPC$_3$ is measured (one after another) in presence of DCB at $T = 18$ mK with the same electromagnetic environment ($R = R_K/2$ and $E_C \approx 25$ μeV). We see that the three QPCs are interchangeable up to $e|V_{\text{dc}}| \approx E_C$. The intrinsic conductance can be extracted from the conductance at large d.c. voltage bias (in the $\pm[45, 51]$ μV gray shaded areas) to obtain the transmission $\tau_{1,2,3} = G^{\text{DCB}}_{1,2,3}(e|V_{\text{dc}}| \gg E_C)/G_K$. **c** These individual transmission probabilities $\tau_{1,2,3}$ are averaged (from six values corresponding to positive/negative bias and QPC$_{1,2,3}$) to give τ with a standard error on the mean value $\Delta\tau$

Influence on a distant gate

The influence on the change on the voltage gates of a QPC$_i$ onto the transmission of a QPC$_j$ is maximal when $\tau_j \sim 0.5$ because the slope $\partial\tau/\partial V_{\text{qpc}}$ is maximal around this value. To be more accurate, we characterize the crosstalk QPC$_i$ $\rightsquigarrow$ QPC$_j$ in two steps. First, we try to adjust the coefficient α such that the variation, $\delta\tau_j$, induced on the transmission of QPC$_j$ by a step δV_{qpc_i} is compensated by $\delta V_{\text{qpc}_j} = \alpha \delta V_{\text{qpc}_i}$ so that we can sweep V_{qpc_i} (on the range $\tau_i : 0 \to 1$) and keep τ_j constant. Second, we determine the absolute shift ΔV_{qpc_j} measured on τ_j when we pass from $\tau_i = 0$ to $\tau_i = \tau_i^{\text{ref}}$ (where τ^{ref} is a picked up value close to half transmission for the QPCs and $\tau_{sw}^{\text{ref}} = 2$ when calibrating the influence of a SW gate onto a QPC).

These calibrations of the crosstalks between the surface gates are done using different configuration of the switch gates. A full set of calibration with their corresponding configuration of all the surface gates is given in Table A.1. There are several remarks to do here: (i) the crosstalks QPC$_i$ $\rightsquigarrow$ QPC$_j$ and QPC$_i$ $\leftrightsquigarrow$ QPC$_j$ are not equal (despite the electrostatic origin of this effect) in particular because the compensation $\delta V_{\text{qpc}_j} = \alpha \delta V_{\text{qpc}_i}$ depends on $\partial\tau_j/\partial V_{\text{qpc}_j}$, (ii) QPC$_3$ has no switch and (iii) these effects are small and only the first order is considered (crosstalks of crosstalks are neglected). The point (ii) is not a problem in principle since calibrating a crosstalk QPC$_i$ $\rightsquigarrow$ QPC$_3$ is possible even if the QPC$_3$ is not in an 'intrinsic configuration'. Indeed, its environment will be fixed to $R_K/2$ thanks to $\tau_{sw_i} = 2$ (whatever

Table A.1 Crosstalks characterization. The sign of the correction should be adapted to compensate the decrease/increase of the considered QPC. Some values can be accessed with different configurations. The first line of the embraces for QPC_1 and QPC_2 correspond to the 'intrinsic' configuration while the second line involves DCB. For QPC_3 both lines of the embraces correspond to DCB configuration but with different environment. The correction in the embraces are similar and we take the mean value for our experiment (except for $QPC_1 \rightsquigarrow QPC_2$, we take $\Delta V_{qpc_2} = 4.0\,\mathrm{mV}$)

	First step (α)						Second step (ΔV_{gate_j})					
crosstalk $i \rightsquigarrow j$	$\delta V_{gate_i}/\delta V_{gate_j}$	τ_{sw1}	τ_{sw2}	τ_1	τ_2	τ_3	ΔV_{gate_j} (mV)	τ_{sw1}	τ_{sw2}	τ_1	τ_2	τ_3
$QPC_1 \rightsquigarrow QPC_2$	0.0035	2	2	$0 \to 1$	τ_2^{ref}	0	$\{$ 0.3 ± 0.1 0.8 ± 0.1	$\{$ 0 2	$\{$ 2 0	0 & τ_1^{ref}	$0 \to 1$	0
$QPC_3 \rightsquigarrow QPC_2$	0.011	0	2	0	τ_2^{ref}	$0 \to 1$	1.0 ± 0.1	0	2	0	$0 \to 1$	0 & τ_3^{ref}
$QPC_2 \rightsquigarrow QPC_1$	0.0065	2	2	τ_1^{ref}	$0 \to 1$	0	$\{$ 0.8 ± 0.1 0.9 ± 0.1	$\{$ 2 0	$\{$ 0 2	$0 \to 1$	0 & τ_2^{ref}	0
$QPC_3 \rightsquigarrow QPC_1$	0.012	2	0	τ_1^{ref}	0	$0 \to 1$	1.5 ± 0.1	2	0	$0 \to 1$	0	0 & τ_3^{ref}
$QPC_1 \rightsquigarrow QPC_3$	0.005	2	0	$0 \to 1$	0	τ_3^{ref}	$\{$ 0.5 ± 0.1 0.75 ± 0.15	$\{$ 2 2	$\{$ 2 0	0 & τ_1^{ref}	0	$0 \to 1$
$QPC_2 \rightsquigarrow QPC_3$	0.025	0	2	0	$0 \to 1$	τ_3^{ref}	$\{$ 3.1 ± 0.15 3.3 ± 0.15	$\{$ 2 0	$\{$ 2 2	0	0 & τ_2^{ref}	$0 \to 1$
$SW_1 \rightsquigarrow QPC_2$							1.5 ± 0.2	0 & 2	0	2	$0 \to 1$	0
$SW_2 \rightsquigarrow QPC_1$							2.55 ± 0.1	0	0 & 2	$0 \to 1$	2	0
$SW_2 \rightsquigarrow QPC_3$							2.3 ± 0.1	0	0 & 2	0	2	$0 \to 1$

$\tau_i \leqslant 2$). In practice we have noticed an effect of $\tau_i : 0 \to 1$ when we try to adjust this correction, but we can compare with QPC_1 and QPC_2 (which show the similar slight change of environment when they are in 'DCB configuration').

Influence of a switch gate on its lateral QPC

This characterization has been used to extract the transmission τ_i of a QPC_i (with $i = L$ or R) in the charge quantization experiment. It consists in taking a trace of QPC_i for the two configurations of SW_i: 'open' and 'closed' (see Fig. A.7b). In the open position, the QPC_i is sensitive to its electronic environment and one should apply a large d.c. voltage bias ($eV \gg E_C$, to avoid any renormalization of the conductance due to dynamical Coulomb blockade). Note that the voltage drop across the QPC_i in the red inset of Fig. A.7b depends on its transmission as $\tau_L/(\tau_L + \tau_R)V$, one should

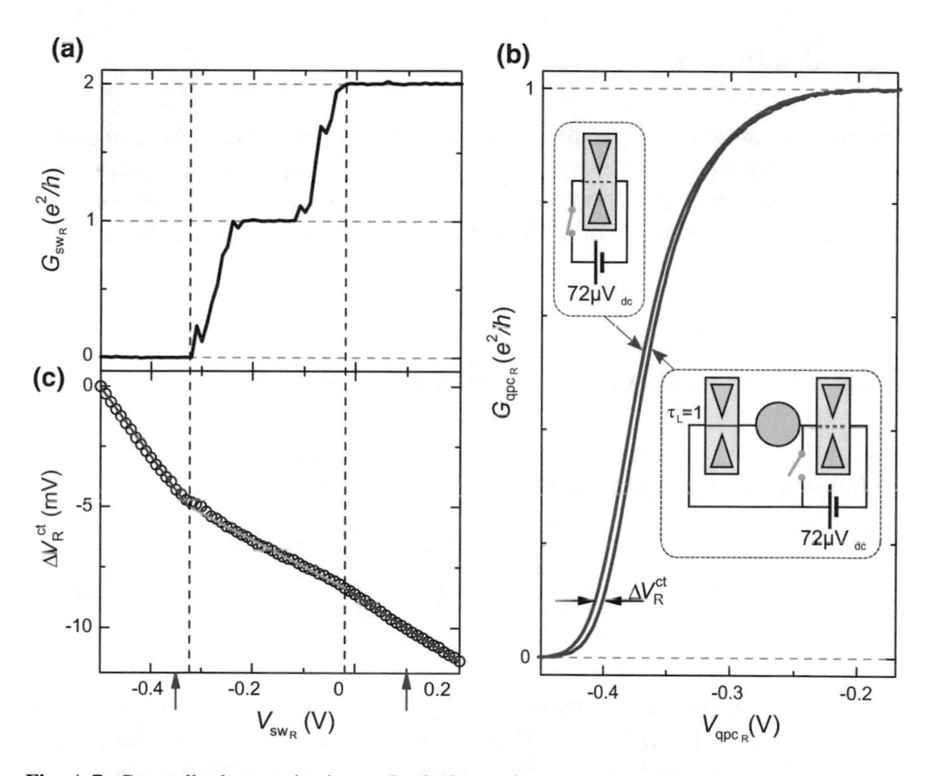

Fig. A.7 Crosstalk characterization. **a** Intrinsic conductance G_{sw_R} across the right switch gate versus the voltage applied on this gate V_{sw_R}. The red and blue arrows on the x-axis indicates the voltages that define respectively the 'open' and 'closed' positions of the switch, see the inset of **b** (a zero-conductance in red is an open circuit). **b** Intrinsic conductance G_{qpc_R} across the QPC_R versus its voltage gate V_{qpc_R}. The red and blue traces correspond respectively to the 'open' and 'closed' positions of the right switch. The capacitive influence is measured by the spacing ΔV_R^{ct} between the two traces (focussing on small conductances, see text). **c** The spacing ΔV_R^{ct} between a reference trace of G_{qpc_R} measured for a $V_{\text{sw}_R} = -0.5$ V and another one is displayed versus V_{sw_R}. The vertical dashed lines delimit the $G_{\text{sw}_R} = 0$ and $2G_K$ quantum Hall plateaus

then focus on $\tau_R \ll 1$ (to be in the same conditions as in the blue inset where the environment is short-circuited).

The spacing ΔV_R^{ct} between the two traces in Fig. A.7b is directly the correction to apply on the right QPC voltage gate V_{qpc_R} when switching the state of its lateral switch. This spacing ΔV_R^{ct} is evaluated by focusing on small G_{qpc_R} in order to minimize the difference between the voltage bias of QPC_R in both configurations.

We have measured the dependence of this spacing for other positions of the switch than the two we have chosen for its 'open' and 'closed' configurations ($V_{\mathrm{sw}_R} = -0.35$ V and 0.1 V respectively). We see in Fig. A.7c that this dependence is relatively simple (as we might expect for an electrostatic effect in this geometry); it is piecewise linear.

A.2 Ohmic Contact

The presence of the central ohmic contact (in purple in Fig. A.1) and its good quality are essential for the experiments reported in this thesis. It has two important features: (i) the coherence of electron is lost when they cross it because the energy level spacing in this metal is completely negligible and (ii) its connection to the 2DEG is perfect at our accuracy.

A.2.1 Mean Energy Level Spacing in the Island

The energy level spacing depends on the dimension of the conductor. It is given by the following formulas (first equations in [3]):

$$\begin{aligned} \delta E_{2D} &= (1/\pi)\frac{\hbar^2\pi^2}{mL^2} \\ \delta E_{3D} &= 1/(3\pi^2 N_{\mathrm{at}})^{1/3}\frac{\hbar^2\pi^2}{mL^2} \end{aligned} \tag{A.1}$$

where N_{at} is the number of atoms, $m = m_e = 9.1 \times 10^{-31}$ kg is the mass of the electrons and $L \sim 1.5\,\mu$m is the typical length of the material. We estimate $N_{\mathrm{at}} \sim (L/a)^3$ with $a \sim 0.1$ or 1 nm as a typical interatomic distance. These numbers give $\delta E_{3D}/k_B \approx 0.1$ or $1\,\mu$K. This is far below our lowest base temperature of 6 mK. At this temperature, too many discrete energy levels will be activated to still consider the density of state as discrete.

The presence of this metallic ohmic contact is crucial, because $\delta E_{2D} \approx 18$ mK (here we should consider the effective mass $m^* = 0.067 m_e$ of the electrons in GaAs). This level spacing (for the 2DEG without ohmic contact) would be larger than our base temperature.

We distinguish our metallic island (that has a continuous density of state) from the *quantum dots* (that have a discrete one) used in other similar experiments as [4, 5]. Having discrete levels would have dramatically changed the physics involved in our device where coherent transport through the island connected by few electronic channels is fully negligible.

A.2.2 *Connection of the Micron-Sized Ohmic Contact to the 2DEG*

At our experimental accuracy, we have not measured any reflection at the interface between the ohmic contact and the 2DEG for the first electronic channel transmitted through each QPC. The measurement protocol is explained here, it is similar to the way we calibrate the gain of the measurement lines in Appendix C.1.2. However, this measurement is independent of any gain calibration. Measurement artifacts could only come from offsets (which are smaller than experimental noise for this characterization).

We measure the signal reflected back $V_{i,i}$ on each QPC when all the QPCs are at $\tau = 0$ and then when they all are at $\tau = 1$. The ratio of the two measurements will be linked to the transmissions by:

$$\nu\left(1 - \frac{V_{i,i}(\tau = 1)}{V_{i,i}(\tau = 0)}\right) = \left(\frac{1}{\tau_i} + \frac{1}{\tau_j + \tau_k}\right)^{-1} \tag{A.2}$$

where ν is the filling factor of the IQHE. Let us define the normalized signals $v_{i,i} \triangleq \nu\left(1 - \frac{V_{i,i}(\tau=1)}{V_{i,i}(\tau=0)}\right)$. We have 3 equations and 3 unknowns, solving the system of equations yields the expression of τ_i:

$$\tau_i = \frac{v_{i,i}^2 + (v_{j,j} - v_{k,k})^2 - 2\, v_{i,i}(v_{j,j} + v_{k,k})}{2(v_{i,i} - v_{j,j} - v_{k,k})} \tag{A.3}$$

The measurements and the deduced transmissions are given in Table A.2.

Table A.2 Transmission of the outermost channel to the micron-sized central ohmic contact. These data are taken at the filling factor $\nu = 3$. Each configuration ($\tau = 0$ or $\tau = 1$) have been acquired during ~20 min. The errors bars are standard error on the mean value

QPC_i	$V_{i,i}(\tau = 0)$ (μV)	$V_{i,i}(\tau = 1)$ (μV)	τ_i
QPC_1	3.82606 ± 0.00002	2.97598 ± 0.00002	0.99967 ± 0.00006
QPC_2	3.79625 ± 0.00002	2.95269 ± 0.00002	0.99991 ± 0.00006
QPC_3	3.80858 ± 0.00002	2.96212 ± 0.00002	1.00031 ± 0.00006

Let us recall that this measurement is free of any gain calibration as we divide two measurements made with the same amplification line. The offsets in the voltage measurement are typically of -7×10^{-11} V. Such an offset will change the transmission measured of $\Delta\tau = 2 \times 10^{-5}$, which is smaller than the measurement error bars. That is why we can claim that the connection between the micron-sized ohmic contact and the 2DEG is perfect within our accuracy.

This connection is weaker for inner channels. It remains good at 10^{-3} for the second channel at $\nu = 3$. The last channel is not transmitted to the central ohmic contact on the sides of QPC_1 and QPC_3, the transmission remains at 2 while it reaches $\tau_2 = 3.00221$ for QPC_2. Despite the imperfection of the connection of the ohmic contact to the 2DEG for these inner channels (probably because of its small size), we are satisfied since we are not using these inner channels in our experiments.

Appendix B
Single Electron Transistor

This appendix is a pedestrian derivation of the conductance of the SET based on the perturbative approach for the tunneling through the junctions explained in [6, 7]. This theory (usually called the 'orthodox theory') is well established. Here we aim to present the origin of quantities we will use all along this thesis (the charging energy E_C, the conductance of the device, etc.). Note that this theory is semi-classical (the only 'quantum' ingredient is the amplitude of the tunneling probability).

B.1 Charging Energy

Let us consider the electrical circuit sketched in Fig. B.1. The central node is surrounded by capacitors. The border of the so-called 'island' is delimited by a dashed line. The total charge Q in the island is an integer multiple of the elementary charge $Q = Ne$.

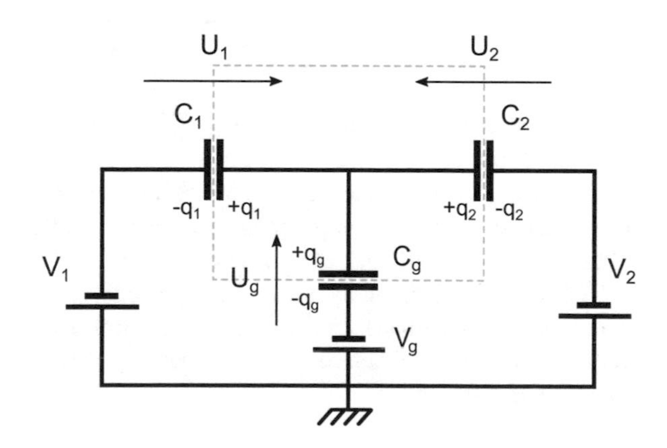

Fig. B.1 Single electron transistor (SET) circuit. The 'island' is delimited by the gray dashed box. Each branch surrounding the central node is labeled with an index $i = 1, 2, g$. U_i indicates the voltage drop across the capacitor i, and $\pm q_i$ is the charge accumulated on each side of the capacitor i

Z. Iftikhar, *Charge Quantization and Kondo Quantum Criticality in Few-Channel Mesoscopic Circuits*, Springer Theses,
https://doi.org/10.1007/978-3-319-94685-6

The electrostatic energy of the island is given by the sum of the energy stored in the capacitors C_1, C_2, C_g and the potential electrostatic energy given by the voltage sources V_1, V_2, V_g to bring by the charges q_1, q_2, q_g onto the island:

$$E_{el} = \frac{1}{2}\left(\frac{q_1^2}{C_1} + \frac{q_2^2}{C_2} + \frac{q_g^2}{C_g}\right) + q_1 V_1 + q_2 V_2 + q_g V_g \tag{B.1}$$

We can simplify this expression using Kirchoff laws and the definition of the charge state $N \triangleq (q_1 + q_2 + q_g)/e$:

$$\begin{cases} U_1 - U_g = V_g - V_1 \\ U_2 - U_g = V_g - V_2 \\ q_1 + q_2 + q_g = eN \end{cases} \tag{B.2}$$

where $U_i = q_i/C_i$ is the voltage drop across the capacitor $i = \{1, 2, g\}$. If we substitute the solution of this system of equation in the Eq. (B.1), we get

$$\begin{aligned} E_{el} = \frac{1}{2(C_1 + C_2 + C_g)} \big[& e^2N^2 + \\ & -C_1(C_2(V_1 - V_2)^2 + C_g(V_1 - V_g)^2) - C_2C_g(V_2 - V_g)^2 \\ & +2eN(C_1V_1 + C_2V_2 + C_gV_g)\big] \end{aligned} \tag{B.3}$$

We are interested in knowing which N will minimize the electrostatic energy E_{el} for a given configuration of voltages V_i. This is why we can add or remove terms that does not depend on N. Let us define

$$q \triangleq -(C_1V_1 + C_2V_2 + C_gV_g) \tag{B.4}$$

By removing the terms independent of N, and adding $q^2/(2(C_1 + C_2 + C_g))$, it comes:

$$\boxed{E_{el} = E_C(N - q/e)^2} \tag{B.5}$$

where we have used the charging energy E_C defined by: $E_C \triangleq e^2/2C_\Sigma$, with $C_\Sigma \triangleq C_1 + C_2 + C_g$.

Unlike q_1, q_2 or q_g, the variable q has the unit of a charge, but it is continuous. The electrostatic energy only depends on this variable, it is then easy to minimize: when $-1/2 \leqslant q/e \leqslant 1/2$, $N = 0$, when $1/2 \leqslant q/e \leqslant 3/2$, $N = 1$, and so on.

The minimal electrostatic energy is e-periodic, and at each half-integer values of q/e there are two adjacent values of N that minimize E_{el}. It means one can tune q to put the system in a degenerate state. Dealing with this degenerate state would be central in this thesis.

B.2 Coulomb Diamonds

So far we have studied a static situation. Let us now define the processes that could change the number of electrons N on the island. We distinguish between the capacitors C_1 and C_2 that could leak electrons by a tunnel process (that will be discussed in the Appendix B.3) and the capacitor C_g that will remain a perfect capacitor.

There are four possible processes: an increment of N by a charge (i) coming from the reservoir of the voltage source V_1 or (ii) coming from V_2, or a decrement of N by a charge (iii) going to the voltage source V_1 or (iv) going to the voltage source V_2. The electrostatic energy before and after these process is not the same.

For instance, let us consider the process (i). Replacing $q_1 \rightarrowtail (q_1 + 1)$ and $N \rightarrowtail (N + 1)$ in the same calculations as in the previous section leads to $E_{el}^{\text{after}} = E_C((N + 1) - q/e)^2$. Here we also have to take into account the work done by the voltage source V_1 to bring a new charge e onto the island. For this process (i), the difference of energy after (i) and before (i) is given by $\Delta E_1^+ = E_{el}(N + 1) - E_{el}(N) + eV_1$, where the superscript '+' symbolize the incrementation of the number of electrons on the island and the subscript '1' indicates the voltage source that moves the electron.

In general, $\Delta E_i^\eta = E_{el}(N + \eta) - E_{el}(N) + \eta e V_i$ with $\eta = \pm 1$, we can simplify this in:

$$\Delta E_i^\eta = 2E_C(\eta N + 1/2 - \eta q/e) + \eta e V_i \tag{B.6}$$

Let us consider the case where $N = 0$, with an anti-symmetric biasing $V/2 \triangleq V_1 = -V_2 \triangleq \varepsilon_i V_i$. And let us also define $N_g \triangleq C_g V_g/e$, so that $q = -((C_1 - C_2)V/2 + eN_g)$. The processes (i-iv) can occur only if $\Delta E_i^\eta > 0$:

$$2E_C\left[1/2 + \eta\left(\frac{(C_1 - C_2)V}{2e} + N_g\right)\right] + \eta\varepsilon_i eV/2 > 0 \tag{B.7}$$

As one can see in Fig. B.2 that the region of the (N_g, eV) plane where the four processes (i-iv) are forbidden is an horizontally stretched diamond delimited[3] by the following equations:

$$eV = \frac{4\eta E_C(1/2 + 2\eta N_g)}{2E_C\frac{C_1 - C_2}{e^2} + \varepsilon_i} \tag{B.8}$$

If $C_1 = C_2$, the diamond is symmetric for N_g positive or negative. Otherwise, the slopes of the diamond give access to the asymmetry between the capacitances C_1 and C_2. But whatever the asymmetry, the total height of the diamond is given by $4E_C$.

Experimentally, the fact that all the processes are forbidden means that the charge state N of the island is frozen. No current can flow through it, this phenomenon is called '(static) Coulomb blockade'. This diamond pattern repeats itself at each integer value of N_g, just as E_{el} is e-periodic. On the y-axis, we see that when $eV \gtrsim E_C$, one leaves the diamond, the *blockade regime*.

---

[3]One can neglect the term $2E_C\frac{C_1-C_2}{e^2} = \frac{C_1-C_2}{C_\Sigma}$, and the sign of the inequalities are given by the sign of $\eta\varepsilon_i$.

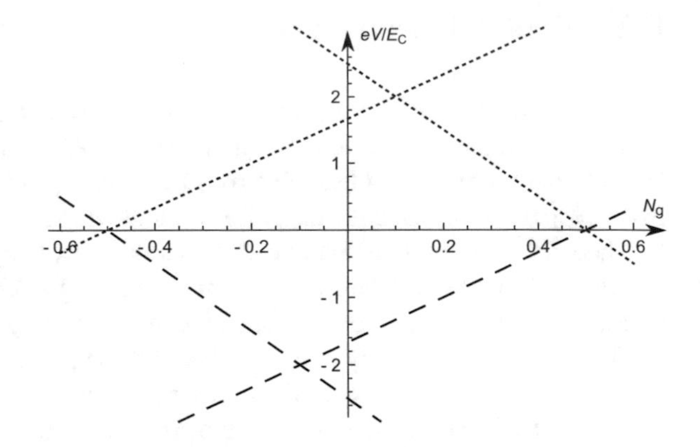

Fig. B.2 Coulomb diamond. The four lines of the diamond are defined by Eq. (B.8). No tunneling process can occur above (below) the dashed (dotted) lines. The conductance of the SET is zero inside the diamond. The total height of the diamond is $4E_C$. In this plot: $e = 1$, $C_1 = 1.0$ and $C_2 = 0.9$

In practice, the experiment will be run at finite temperature. This will smear the diamonds and allow a current to flow in the vicinity of the lines defined by Eq. (B.8). To be more quantitative, we should introduce the master equation that rules the statistic thermal population of each charge state.

B.3 Master Equation

The processes that change the charge state of the island are due to tunneling of electrons. These events happen randomly when the energy is available.[4] Given N_{states} possible charge states, we want to know which charge state is populated in given operating conditions.

The evolution of the probability $p_N(t)$ for the island to have N electrons at time t is ruled by the following master equation:

$$\begin{aligned}\frac{\mathrm{d}}{\mathrm{d}t} p_N(t) = &- \left(\Gamma^+(N) + \Gamma^-(N)\right) p_N(t) \\ &+ \Gamma^-(N+1) p_{N+1}(t) + \Gamma^+(N-1) p_{N-1}(t)\end{aligned} \tag{B.9}$$

where the Γ are tunneling rates. These tunneling rates depend on the working conditions, for instance, if the island is in charge state N and it is easy to go to charge state $N + 1$, the rate $\Gamma^+(N)$ will be high. $1/\Gamma$ is the typical time for a process to occur.

In the Eq. (B.9), the two first terms with sign '$-$' stands for the processes that depopulates the charge state N, while the two last terms stands for the processes that populates this state. Each rate Γ of Eq. (B.9) contains both the contributions for tunneling process from/to the reservoir of the voltage sources V_1 and V_2 : $\Gamma^\eta = \Gamma^\eta_1 + \Gamma^\eta_2$.

[4]We do not consider *co-tunneling*: the tunneling of an electron to a virtual state (which is higher in energy) and then, from this virtual state to a real state.

The current leaking through the capacitor C_1, or in more accurate words, the current flowing through the tunnel junction #1 is e times the net number of charges that go through the junction per unit of time:

$$I_1 = e \sum_{N_{\text{states}}} \left(\Gamma_1^+(N) - \Gamma_1^-(N)\right) p_N \tag{B.10}$$

We assume the stationary hypothesis: $\frac{\mathrm{d}}{\mathrm{d}t} p_N(t) = 0$. In this case, there is no charge temporary stored in the island, and the current $I \triangleq I_1 = I_2$ through the two junctions is the same. This hypothesis is valid since the tunneling rates Γ are very fast compared to any change in the working conditions. Indeed, the tunneling rates are given by:

$$\Gamma = \frac{G_\infty}{e^2} \frac{\Delta E}{\exp(\Delta E / k_B T) - 1} \tag{B.11}$$

where $G_\infty = \tau e^2 / h$ and τ is the intrinsic transmission of the junction. At small bias V, and close to the charge degeneracy $N_g \sim 1/2$, $\Gamma \sim \tau k_B T / h$. With a typical value of $\tau = 0.10$ and a temperature of $T = 6\,\text{mK}$, $1/\Gamma \sim 10^{-7}$sec. This time is much smaller than the frequencies we use in our measurement which are in the kHz range or even lower.

B.4 Coulomb Blockade Oscillations

Now we have all the ingredients to trace the electrical conductance of the device as a function of N_g (which is proportional to the gate voltage V_g, $N_g \triangleq C_g V_g / e$). The conductance will show e-periodic oscillations with maxima placed at the charge degeneracy points.

Let us consider only two charge states $N = 0$ or 1. One can solve the master equation Eq. (B.9) and use $p_0 + p_1 = 1$ to get p_0 and p_1. Using Eq. (B.10), it comes:

$$I/e = \frac{\Gamma_1^+(0)\Gamma_2^-(1) - \Gamma_1^-(1)\Gamma_2^+(0)}{\Gamma^+(0) + \Gamma^-(1)} \tag{B.12}$$

And using Eqs. (B.11) and (B.6), we get the explicit expression of $I(G_{1\infty}, G_{2\infty}, T, V_1, V_2, q(V_1, V_2, V_g))$. If we assume a symmetric small biasing $V/2 \triangleq V_1 = -V_2$, the first term in the series development of the conductance $G = I/V$ is:

$$\boxed{G = \frac{G_\infty}{2} \frac{\Delta E / (k_B T)}{\sinh \Delta E / (k_B T)}} \tag{B.13}$$

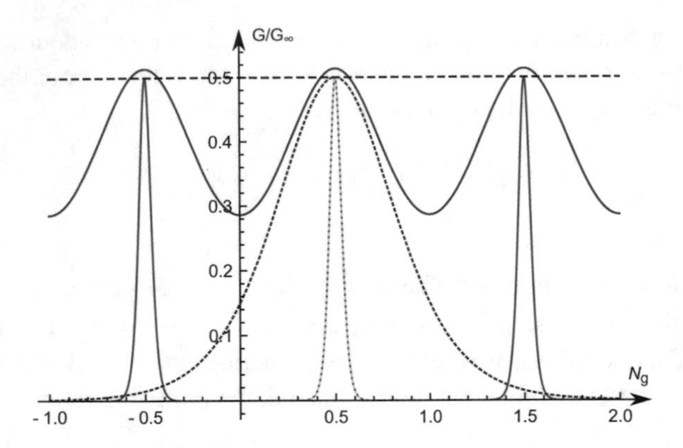

Fig. B.3 Coulomb blockade oscillations of conductance. The conductance of the SET as a function of the plunger gate N_g is plotted for two models (solid and dotted lines) and for two temperatures (blue-cyan at 10 mK and red-violet at 100 mK) with $E_C = 300\,\text{mK}$. The dotted lines are plotted taking into account only two charge states, while the solid ones take $N_{\text{states}} = 7$. The solid lines model is *not* given by an accumulation of independent dotted lines. However, the higher the temperature, the higher the conductance

where $G_\infty \triangleq \dfrac{G_{1\infty} G_{2\infty}}{G_{1\infty} + G_{2\infty}}$ is the intrinsic serial conductance and $\Delta E \triangleq 2E_C(1/2 - N_g)$ is related to the gate voltage V_g (and which will be called the 'splitting in energy' in Chap. 4, because it favors either the $N = 0$ or the $N = 1$ state, while at $w = 0$ these two states are degenerate in energy).

Actually, we should consider more charge states to obtain e-periodic peaks. Figure B.3 shows the normalized conductance G/G_∞ as a function of $q/e \propto N_g$ for a small biasing and for two temperatures. In this figure, we see that at higher temperatures, one has to consider many charge states in the calculations, because the thermal distribution populates several charge states. At very high temperatures, the oscillations vanishes: the charging effects are no longer visible, as the electrons are too energetic to be Coulomb blocked.

These Coulomb oscillations are a clear signature of the quantization of the charge on the island. Indeed, the e-periodic dependence of the conductance of the device on the gate voltage V_g shows a precise control of the number of electrons *located* on the island.

Appendix C
Experimental Procedures

In the first section of this appendix, we explain the method we used to extract the conductance of the whole device from voltage measurements. The in situ conductance of a single QPC can also be extracted when none of the three QPCs is completely pinched off.

The second section discusses the tuning of the device to observe 'charge' Kondo effect and the methods used to extract the conductance at energy level degeneracy. In the last section we explain the systematic treatment of large sets of measurements and we discuss some experimental artifacts to handle.

C.1 Extraction of in situ Conductances

Irrespectively of the configuration of the QPCs and other surface gates, the resistance of the sample[5] is fixed by IQHE to $1/(\nu G_K)$. We use lock-in amplifiers to measure the voltages defined in Fig. C.1. At first, we will assume that the voltage are measured without any calibration offset and that all the measurement lines are equivalent (same gain). In the second subsection, we will explain how to calibrate the gain and the offsets.

C.1.1 Conductance Formulae

We assume that the electronic channels are completely reflected on the characterization gates ($\tau_{sw_1} = \tau_{sw_2} = 0$, uncolored in Fig. C.1) and that only the outermost

[5] One should distinguish the conductance of the *sample*, which means 'between a given input port of the circuit and the ground', from the conductance of the *device*, which means 'between an input port and an output port'. The latter depends on the configuration of the surface gates.

Z. Iftikhar, *Charge Quantization and Kondo Quantum Criticality in Few-Channel Mesoscopic Circuits*, Springer Theses,
https://doi.org/10.1007/978-3-319-94685-6

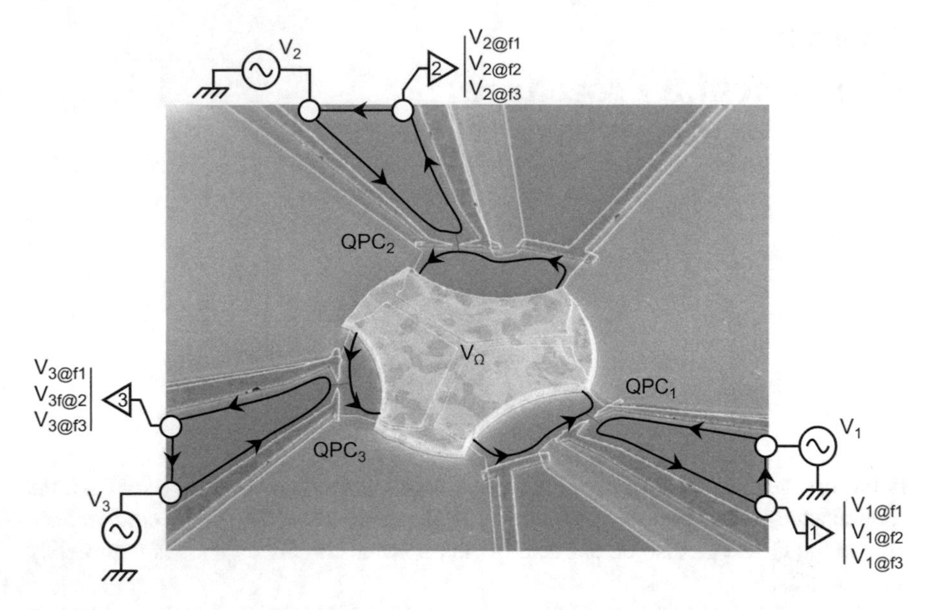

Fig. C.1 Measurement schematics. We use three low frequency a.c. voltage sources (labeled V_j) to inject the signal in the IQHE edge channels (black tracks). Lock-in amplifiers are used to measure simultaneously nine voltages a.c. signals. The voltage measured at the point i in response to the voltage V_j injected "at frequency j" is labeled $V_{i@fj}$

channels have partial transmission $\tau_i \leq 1$ (otherwise the formulae would be different). As the current is conserved, the injected voltage V_j is the sum of the voltages measured "at frequency j":

$$V_j = \sum_{i=1}^{3} V_{i@fj}$$

There are basically two useful formulas. Either one considers the current transmitted through the island (voltages as $V_{i@fj}$, with $i \neq j$) or the current reflected at QPC$_i$ (the voltage $V_{i@fi}$).

Applying the Kirchoff's current law successively on contact i and on the central ohmic contact Ω yields:

$$\nu G_K V_{i@fj} = G_i V_{\Omega@fj} = \frac{G_i G_j}{G_1 + G_2 + G_3} V_j \tag{C.1}$$

where $V_{\Omega@fj}$ is the voltage on the central ohmic contact measured at the frequency of the source V_j. To get the formula for the current in reflection, we just add the contribution of the inner quantum Hall channels:

$$\nu G_K V_{i@fi} = G_i V_{\Omega@fi} + (\nu G_K - G_i) V_i = \left(\nu G_K - G_i \left(1 - \frac{G_i}{G_1 + G_2 + G_3}\right)\right) V_i \tag{C.2}$$

We can therefore extract the in situ conductances G_i of each QPC from the voltage measurements. Measurements of reflected and transmitted signals are redundant, because of current conservation. Thus we define the following "rescaled voltages" G_{ii} (that have the dimension of a conductance) and that allow for a direct averaging of the reflected and transmitted signals:

$$G_{ii} \triangleq \nu G_K \left(1 - \frac{V_{i@fi}}{V_i}\right) = \nu G_K \frac{\sum_{j\neq i} V_{j@fi}}{V_i}$$

This measured conductance G_{ii} depends on the in situ conductances as:

$$G_{ii} = \frac{\sum_{j\neq i} G_j G_i}{G_1 + G_2 + G_3} \tag{C.3}$$

Once we have measured the G_{ii} for all $i = \{1, 2, 3\}$, we can invert the system and find the in situ conductances G_1, G_2 and G_3. The single assumption required to extract the in situ conductances is the validity of the Kirchoff's law on each contact of the nanofabricated sample.

Note that if one of the three QPCs is completely pinched off (say $G_k = 0$), we can only access to the serial conductance $G_i G_j/(G_i + G_j)$ and cannot extract the individual conductances G_i and G_j.

C.1.2 Calibration of the Gain and the Offset

We use the same kind of voltage sources, pre-amplifiers and lock-in amplifiers in all the measurement setup. However, in practice, the gain on each measurement line may slightly differ and a systematic offset can exist.

Offset

In order to calibrate the offset, one just needs to measure all the voltages $V_{i@fj}$ for a plugged and unplugged voltage source. This is done with all the QPCs completely pinched off, therefore, by this most accurate procedure, we cannot calibrate the offset on the $V_{i@fi}$ measurements (in reflection). We found the same typical offset of $\sim -7 \times 10^{-11}$ V on all the voltages $V_{i@fj}$ (with $i \neq j$).

For comparison, the voltage applied on the sample at base temperature $T \approx$ 7.9 mK is $V_{\text{a.c.}} \approx 4.8 \times 10^{-7}$ V (which is of the order of the thermal energy that corresponds to a voltage of $k_B T/e \approx 6.8 \times 10^{-7}$ V). This offset becomes important when measuring weak signals, when one of the conductances is very small (see Fig. 4.10).

Gain

In Table A.2 we have shown that, on the first quantum Hall plateau, the transmission of the outermost channel to the central ohmic contact is perfect within our accuracy.

That measurement was independent of any gain calibration because we used ratio of voltages from the same measurement line (in reflection $V_{i@fi}$).

In order to calibrate the gain of the measurement lines, we set all the QPCs to transmission $\tau_1 = \tau_2 = \tau_3 = 1$. We assume these transmissions to be exactly equal to one as previously checked. Note that for this calibration, although the calibrated offsets can be taken into account, they are negligible since the signals are relatively large. In this configuration, the ratio of measured voltages (e.g. $V_{1@f3}/V_{2@f3}$) directly give ratio of gains (e.g. γ_1/γ_2 at "frequency 3", where γ_i is the gain on the i-th measurement line).

We noticed that, the gains $\gamma_1/\gamma_2 = 1.0018$ and $\gamma_3/\gamma_2 = 1.0095$ (with $\gamma_2 = 1$ taken as reference) yield the same G_{ii} for both the measurements in reflection and in transmission: $(1 - V_{i@fj}/V_i)/((V_{j@fi} + V_{k@fi})/V_i) = 1 \pm 5 \times 10^{-5}$. When considering our three frequencies $f_1 = 145$ Hz, $f_1 = 163$ Hz and $f_1 = 185$ Hz, the typical variation of a gain ratio with frequency ($\sim 4 \times 10^{-3}$) is within the uncertainty bar of the gain calibration.

C.2 Device Tuning to Measure the Multi-channel 'Charge' Kondo Effect

In order to observe the flow of the conductance towards the N-channel 'charge' Kondo fixed point, one needs to set symmetric transmission $\tau_1 = \tau_2 = ... = \tau_N$ of N electronic channels and also to tune the voltage gate V_g at charge degeneracy $\delta V_g = 0$.

C.2.1 Symmetric Coupling

Our goal is to know the voltage to apply on each QPC to have the same transmission $\tau \triangleq \tau_1 = \tau_2 = \tau_3$, and we want a dozen of values picked in the range $\tau \in [0, 1]$.

Approximate tuning

The first step is to use the characterization gates to measure the intrinsic transmission (in DCB regime at large voltage bias). Then we get the voltage $V_{\text{qpc}_{i,k}}$ to apply on QPC$_i$ to get $\tau_{i,k}$ (where $\tau_{i,k}$ refers to the k-th picked up value on QPC$_i$).

We know the correction to apply on V_{qpc_i} to go from the DCB regime to the configuration where all the switches are at $\tau_{\text{sw}} = 0$ (see Appendix A.1.4). We have determined each $V_{\text{qpc}_{i,k}}$ while the other QPCs were in a reference configuration ($\tau_{j \neq i} = 0$), but we also know the correction to apply on V_{qpc_i} when we change the voltage on the other QPCs.

It means that we are able to measure Coulomb oscillations with all the QPCs having the same transmission: $\tau_1 = \tau_2 = \tau_3$.

Fine tuning

When none of the three QPCs is completely pinched off ($\tau_{1,2,3} \neq 0$), we are able to find the three individual conductances G_i. We want to verify whether the in situ conductances are symmetric $G_1 = G_2 = G_3$ for the multi-channel Kondo experiment. In this case, there are Coulomb oscillations and we focus on maxima of the conductances $G_i(\delta V_g \approx 0)$ where the signal is maximal.

The procedure explained in the previous subsection relied on several capacitive crosstalk calibrations which limits our accuracy. We therefore perform an additional fine adjustment of the QPC symmetry.

For this fine symmetry tuning, we use QPC_2 as a reference at the temperature $T \approx 18\,\mathrm{mK}$, and we adjusted the two others to have all the in-situ conductances exactly symmetric. After this fine tuning we have noticed that the values of $V_{\mathrm{qpc}_{i,k}}$ have been shifted by $\sim 0.2\,\mathrm{mV}$ for QPC_1 and by $\sim 0.4\,\mathrm{mV}$ for QPC_3. These shifts are compatible with the uncertainty intervals on the crosstalk calibrations.

Approximate transmission symmetry requirement

The effect of an asymmetry between the transmissions is visible on the renormalization flow diagram shown in Fig. 4.9 where we used the transmissions τ characterized in Fig. A.6. The difference between two successive transmissions ranges from 0.1 down to 0.03 in the vicinity of the 3CK fixed point where the step is finer. Even such a small difference has a dramatic effect on the conductance renormalization flow (see the criticality of the symmetry on the first off-diagonal arrows of Fig. 4.9). The requirement on the transmission symmetry to flow towards the right fixed point is therefore more restrictive than $\delta\tau < 0.03$.

C.2.2 Transmissions in Practice

Dependence in energy $\tau(E)$

We tried to minimize this dependence by choosing the best operating points for the QPCs (e.g. see Fig. 1.5). The relative deviation $\Delta\tau/\tau$ between the transmission at equilibrium and the transmission at large voltage bias is typically of $\approx 10\%$, it is maximal in the tunnel case $\tau \ll 1$ (see Fig. A.5).

Stability of the QPCs

The data acquisition of the multi-channel Kondo effect measurement took several months (because of many tests, calibrations, temperature measurements and other time consuming activities). We are able to correct for rare and small shifts of the QPCs. Before all our Kondo effect measurements (that take typically a night or a weekend), we perform a fine measurement of the intrinsic transmission $\tau(V_{\mathrm{qpc}})$ of each QPC and adjust for a possible small V_{qpc} shift with respect to a reference measurement. And after the measurement, we redo a fine measurement of the transmissions to check if the QPCs were stable during the acquisition (if they were not, we redo the measurement).

C.2.3 Tuning to Charge Degeneracy $\delta V_g = 0$

In order to increase our precision on the conductance at charge degeneracy $G(\delta V_g = 0)$, we fit the conductance peak to a model. However, there is no prediction for the general shape of the conductance peak at arbitrary transmissions τ_i and temperature T.

For instance, in the two-channel case, theory predict a shape as $f(V_g) \propto (\alpha V_g/k_B T)/\sinh(\alpha V_g/k_B T)$ in the tunnel regime $\tau \ll 1$ and a sinusoidal shape in the strong coupling regime $1 - \tau \ll 1$. Therefore, we use a general model and extract $G(\delta V_g = 0)$ in several steps to avoid any systematic error due to the specificity of the fitting function. Here follows the procedure used for the 3CK measurements.

Raw data and approximate tuning
In the 3CK measurement, we have three conductances to extract (whereas in 2CK we have only one). We apply no voltage bias V_{dc}, the maximum of the conductance appears therefore at the same gate voltage $\delta V_g = 0$ for all the signals. Thus we start with averaging the three signals $G_{avg} \triangleq (G_{11} + G_{22} + G_{33})/3$, an example is shown in Fig. C.2. The peaks are periodic and we can extract the highest value of each period. We use these maxima as a starting position for the fits.

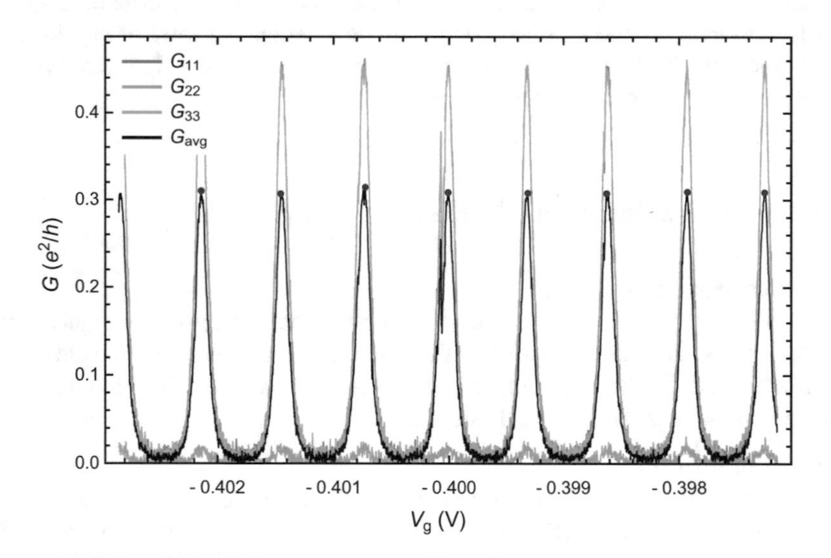

Fig. C.2 Raw experimental data. The conductance G_{ii} are plotted versus the gate voltage V_g at $T \approx 7.9$ mK for $\tau_1 \approx \tau_3 \approx 0.1$ and $\tau_2 \approx 0.79$. Because of the symmetry $\tau_1 \approx \tau_3$, the signals G_{11} and G_{33} are superimposed. The black trace G_{avg} correspond to the average of the three signals. The red dots show the local maxima of G_{avg}

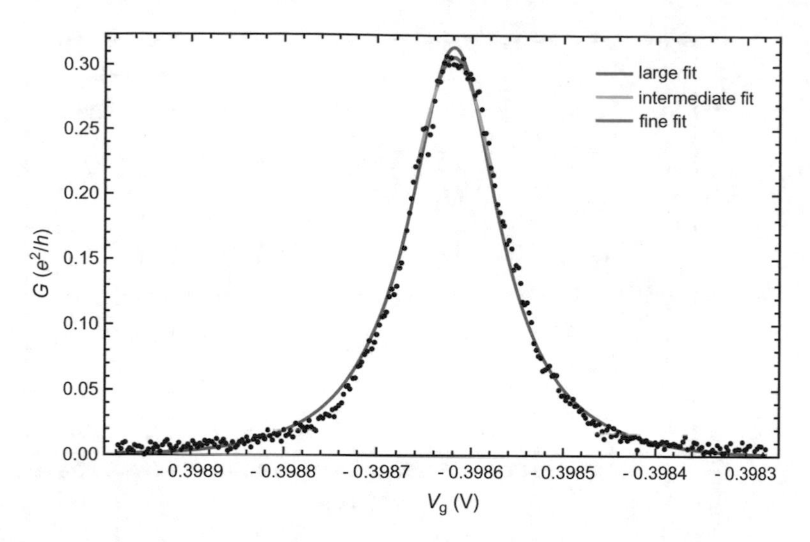

Fig. C.3 Fit of experimental data in three steps. We focus on the third peak (from the right) of Fig. C.2. The three successive fits are performed on different ranges (colored traces)

Large, intermediate and fine fit of the averaged data
In this subsection, we are only interested on the position of the maxima of the conductance. This procedure is applied to each peak of G_{avg} individually. We will determine the value of the conductance maxima $G_{ii}(\delta V_g = 0)$ in the next subsection.

We use the following model based on the Airy function and that have four parameters (an offset m, an amplitude Δ, a finesse F and an origin x_0):

$$m^2 + \Delta^2 \left(1 - \frac{(1 + F^2)\sin^2(k(x - x_0))}{1 + F^2 \sin^2(k(x - x_0))}\right) \tag{C.4}$$

where x is the variable and k is replaced by $\pi/\overline{\Delta x}$, where $\overline{\Delta x}$ is the average distance between the peaks (which is proportional to $1/C_g$ the capacitance of the plunger gate, in practice $\Delta V_g = 0.70\,\text{mV}$). The parameter F gives the finesse of the peak (the model is sinusoidal when $F = 0$ and sharp when F is large).

We will fit the experimental data in three steps. A 'large fit' is used to get an estimation of the parameters. A 'intermediate fit' will be performed on a smaller interval near the maximum to find its position ($\delta V_g = 0$). And finally, we use a 'fine fit' on a tiny region with only Δ as parameter. This last fit gives the value of the conductance at degeneracy independently of the model (since it has been done on a tiny region near the maximum, see Fig. C.3).

The offset m is no longer a parameter in the second fit (called 'intermediate'). Indeed, as the fit is done on a narrower range, the minimum of the oscillations could not be fitted. We set it to the mean value of the set of m found at first fit.

Similarly, the finesse F is not fitted in the 'fine fit' because letting F free when fitting on a tiny region is not a good idea since the global curvature of the peak is not

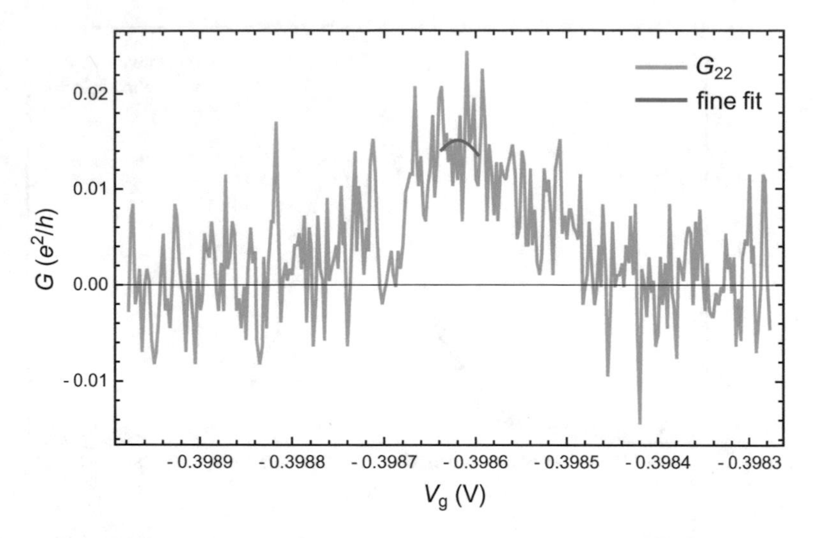

Fig. C.4 Fine fit of a weak signal. The 'fine fit' of the weakest signal of Fig. C.2 (G_{22}, in orange) is shown with the corresponding experimental data

obvious. In this case, we use the median value rather than the mean to avoid wrong values that may happen sometimes because of experimental artifacts (see Fig. C.6 in Appendix C.3).

Conductances at $\delta \mathbf{V}_{g=0}$

The last fit (the 'fine' one) yields the value of the conductance at degeneracy. We are not interested in this value for G_{avg}. However, we know the position of the degeneracy point $\delta V_g = 0$ (which is given by the parameter x_0) for each peak.

To get the individual conductances, we fit again each individual signal G_{ii} using only the 'large' and the 'fine' fits because we just need m and Δ (see Fig. C.4). Indeed, F is essentially the same for the three signals and the 'fine' fit is made on a narrow δV_g interval such that the result does not depend on this fitting parameter. Finally we deduce the in situ conductances G_i at degeneracy from the value of G_{ii} using Eq. (C.3).

Precision and accuracy

The precision on the determination of $G_{ii}(\delta V_g = 0)$ depends on the range over which the fit is performed. This precision increases as $\sqrt{N_{points}}$, where N_{points} is the number of points in the 'fine fit' range. We choose a range that is a function of the finesse F (the wider the peak, the larger the fit range). This has been adjusted to have 'fine fit' ranges small enough, not to be sensitive to the fitting function Eq. (C.4).

In order to increase the accuracy on $G_{ii}(\delta V_g = 0)$, one has to acquire more peaks of a given configuration of transmissions and temperature (e.g. by averaging on all the peaks measured in Fig. C.3). However, one should be careful with respect to some experimental artifacts, as we will see in the next section.

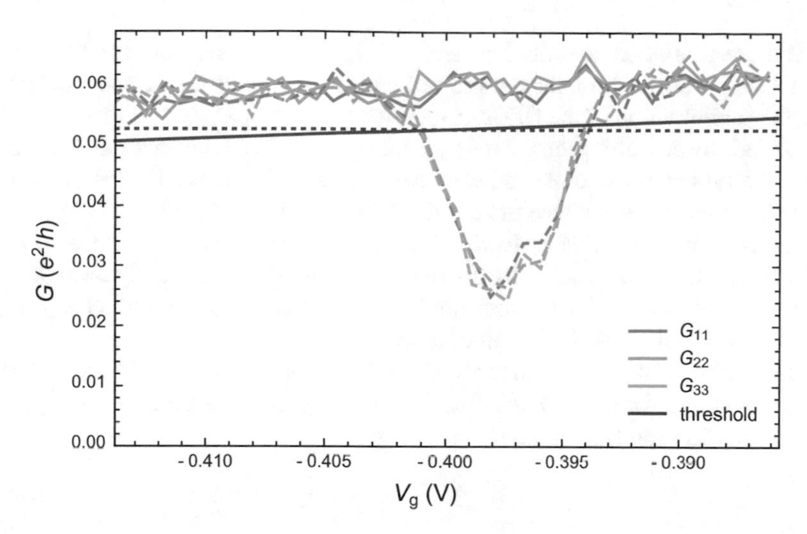

Fig. C.5 Experiment artifact. The conductance at degeneracy $G_{ii}(\delta V_g = 0)$ is plotted as a function of V_g at $T \approx 12$ mK and for a symmetric tunnel configuration $\tau_1 = \tau_2 = \tau_3 \approx 0.1$. Note that each sweep contains about 40 peak, only the conductance at degeneracy is plotted. The conditions were the same for all the curves, indeed the measurement program has just looped back to the first configuration between the evening (solid lines) and the morning (dashed lines). A systematic method excludes the points below the threshold given by the red line (see text in Appendix C.3.1 for the dotted line, and Appendix C.3.2 for the solid line)

C.3 Experimental Artifacts

C.3.1 Systematic Peak Selection

A typical example

Our SET-like device is sensitive to the charge of the environment. We have noticed that some irregular pattern can appear on the Coulomb oscillations. When the transmissions and the temperature are fixed, sweeping the plunger gate V_g should give periodic oscillations with a constant amplitude. In Fig. C.5, we show a typical artifact: a dip in the amplitude of the conductance peak around a specific gate voltage $V_g \approx -0.398$ V.

The known mechanisms (e.g. charge fluctuations) for such artifact can only reduce the conductance. Therefore, we performed a statistical study and exclude the peaks that are anomalously low (this is also true for the 3CK measurements above the 3CK fixed point).

Systematic exclusion

The procedure to exclude the pathological peaks has been automatized because of the large quantity of data (and also to avoid a bias due to human intervention).

At a given temperature, the injected a.c. signal to measure the conductances is fixed, thus the noise σ_G on the normalized measured voltage G_{ii} is also fixed (whatever the configuration of the QPCs or the plunger gate V_g).

For each sweep of V_g (at τ fixed) we have a set of conductances at degeneracy that has a typical noise of σ_G. In each set, we exclude the points that are $n \times \sigma_G$ below the maximal value (see the red dotted line in Fig. C.5, which is plotted with $n = 6$). Assuming a gaussian distribution of the points, this procedure with $n = 6$ excludes less than 1% of normally distributed points (i.e. it essentially excludes only anomalous data). Note that the threshold $n \times \sigma_G$ should be weighted with $\sqrt{N_{\text{points}}}$, the number of points used in the fit of a peak.

About 6 000 peaks have been analyzed to display the data shown in Chap. 3 (symmetric couplings $\Delta\tau = 0$), which means 27 peaks in average for each configuration (of transmission and temperature).

C.3.2 *Crosstalk of the Plunger Gate*

In the 2 and 3CK measurements, we have used V_{sw2} as plunger gate. This voltage was swept around a value where we are sure that $\tau_{\text{sw2}} = 0$ on a typical span of $V_{\text{sw2}} \in [-0.405\,\text{V}, -0.395\,\text{V}]$.

Although this span is very small, we can correct the influence of this sweep on the transmission τ_i of the QPCs during oscillation measurement. If V_{sw2} is crescent (going to positive values), the transmissions will be slightly higher at the end of sweep than at its beginning. This is visible on the conductances shown in Fig. C.5.

We have observed and calibrated this effect at high temperatures (at low temperature, as in Fig. C.5, the signal-to-noise ratio is not high enough to precisely calibrate such a small influence). We have averaged the slopes measured at $T \approx 55$, 80 and 100 mK and used the averaged slope to correct the data measured at all the temperature. One can compare the red solid line threshold in Fig. C.5 that takes into account the crosstalk correction (which is calibrated at $V_{g_0} = -0.4\,\text{V}$) with the red dotted line which is plotted without correction.

C.3.3 *Averaging the Selected Peaks*

In Chap. 4, we study the conductance out of degeneracy $\delta V_g \neq 0$. The peaks we presented are averaged only on the data selected by the systematic procedure described above, in Appendix C.3.1.

Sometimes, when sweeping the plunger gate V_g, there is a jump in the conductance Coulomb oscillations (see Fig. C.6). We eliminate manually the rare affected peaks.

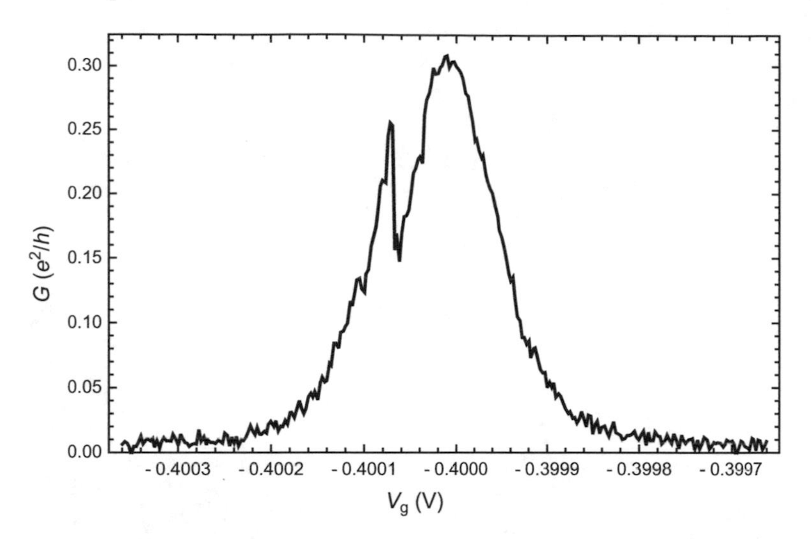

Fig. C.6 Instability of the charge on the island. This instability has been measured on the fourth peak (from the left) displayed in Fig. C.2

References

1. A.K. Mitchell, L.A. Landau, L. Fritz, E. Sela, Universality and scaling in a charge two-channel Kondo device. Phys. Rev. Lett. **116**(15) (2016)
2. Z. Iftikhar, S. Jezouin, A. Anthore, U. Gennser, F.D. Parmentier, A. Cavanna, F. Pierre, Two-channel Kondo effect and renormalization flow with macroscopic quantum charge states. Nature **526**, 233–236 (2015)
3. L.P. Kouwenhoven, C.M. Marcus, P.L. McEuen, S. Tarucha, R.M. Westervelt, N.S. Wingreen, Electron transport in quantum dots. Kluwer Series, editor, Mesoscopic Electron Transport **E345** (1997), pp. 105–214
4. L.P. Kouwenhoven, N.C. van der Vaart, A.T. Johnson, W. Kool, C.J.P.M. Harmans, J.G. Williamson, A.A.M. Staring, C.T. Foxon, Single electron charging effects in semiconductor quantum dots. Zeitschrift für Phys. B Condens. Matter **85**(3), 367–373 (1991)
5. S. Amasha, I.G. Rau, M. Grobis, R.M. Potok, H. Shtrikman, D. Goldhaber-Gordon, Coulomb blockade in an open quantum dot. Phys. Rev. Lett. **107**(21), 216804 (2011)
6. Y.V. Nazarov, Y.M. Blanter,*Quantum Transport: Introduction To Nanoscience* (Cambridge University Press, 2009)
7. G.-L. Ingold, Y.V. Nazarov, Charge tunneling rates in ultrasmall junctions, in *Single Charge Tunneling* (Springer, 1992), pp. 21–107

Printed in the United States
By Bookmasters